# project GMAT
*statistics, permutations & combinations, and probability*

**WRITTEN BY**
**DENIS SOSYURA**

**EDITED BY**
**CHAD TROUTWINE**
**MARKUS MOBERG**

*A comprehensive treatment of the most advanced topics*
*on the quantitative portion of the GMAT.*

ISBN: 0-9761937-0-1

# TABLE OF CONTENTS

**INTRODUCTION** .................................................................................................................. **4**

    *A New Focus* ......................................................................................... 4
    *High Scores* ......................................................................................... 4
    *Beyond the GMAT* ................................................................................. 4

**STRUCTURE OF THE BOOK** ................................................................................................ **5**

**STATISTICS** .......................................................................................................................... **6**

    *Statistics on the GMAT – Recent Trends* ................................................ 6
    *The Good Old Mean* ............................................................................. 6
    MEDIAN, MODE, AND RANGE .................................................................. 12
    STANDARD DEVIATION ............................................................................ 17
    THE MAP STRATEGY FOR STANDARD DEVIATION ...................................... 24
    CHAPTER SUMMARY .............................................................................. 25

**STATISTICS PROBLEM SET** ............................................................................................ **27**

**COMBINATORICS** ............................................................................................................. **42**

    COMBINATORICS MADE EASY .................................................................. 42
    WHEN DOES THE ORDER MATTER? .......................................................... 42
        *Permutations* ................................................................................. 42
        *Combinations* ................................................................................. 43
    PERMUTATIONS: THE WORLD IN ORDER .................................................. 44
    FACTORIALS! ......................................................................................... 45
    PERMUTATIONS WITH REPEATING ELEMENTS .......................................... 50
    PERMUTATIONS INVOLVING SELECTION .................................................... 54
    COMBINATIONS: IN THE WORLD OF ANARCHY .......................................... 60
    ADVANCED TOPICS IN COMBINATORICS .................................................... 65
    STRATEGIES UNIQUE TO COMBINATORICS ................................................ 72
        *1. Permutations or Combinations? Sequence it!* .............................. 72
        *2. Sketch it!* .................................................................................. 72
        *3. Combinatorics Decision Tree* ...................................................... 72
    CHAPTER SUMMARY .............................................................................. 74

**COMBINATORICS PROBLEM SET** ................................................................................. **76**

**PROBABILITY: BEATING THE ODDS** ............................................................................ **89**

    *How It All Started* ................................................................................. 89
    *Concept of Probability and Range of Probability Values* ......................... 89
    PROBABILITY OF A SINGLE EVENT .......................................................... 90
    MUTUALLY EXCLUSIVE EVENTS .............................................................. 95
    COMPLEMENTARY EVENTS ...................................................................... 96
    PROBABILITY OF ONE EVENT <u>OR</u> ANOTHER .......................................... 97
        *1. General Case* ............................................................................. 97
        *2. Mutually Exclusive Events* .......................................................... 102
        *3. Complementary Events* ............................................................... 107
    PROBABILITY OF ONE EVENT <u>AND</u> ANOTHER ...................................... 110
        *1. Dependent Events and Conditional Probability* .............................. 110
        *2. Independent Events* .................................................................... 116
    STRATEGIES UNIQUE TO PROBABILITY .................................................... 122
        *1. The Lucky Twins Strategy: Identifying Pairs in the Answer Choices* ... 122
        *2. The One Mirror Strategy: Constructing Complementary Events* ......... 124
        *3. The Fresh Start Strategy: Differentiating Between Dependent and Independent Events* ... 128
    CHAPTER SUMMARY .............................................................................. 132

STATISTICS

**PROBABILITY PROBLEM SET** ............................................................................................. **135**

**STRATEGIES FOR DATA SUFFICIENCY**................................................................................ **150**

*Recent Trends in Data Sufficiency* ........................................................................... 150

STRATEGIES AND TIPS ........................................................................................................ 150

*1. Start Easy*................................................................................................................. 150

*2. Be BAD* .................................................................................................................... 153

*3. No News Is Good News* ........................................................................................... 156

*4. Memory Blackout* ................................................................................................... 158

*5. Probability Ratios − As Good As It Gets*............................................................... 161

STRATEGY SUMMARY ........................................................................................................ 162

**DATA SUFFICIENCY PROBLEM SET** ................................................................................... **163**

**COMPREHENSIVE PROBLEM SET**....................................................................................... **178**

**CONCLUDING ADVICE**........................................................................................................ **216**

**VERITAS PREP**....................................................................................................................... **217**

*Elite Test Preparation*............................................................................................... 217

**VERITAS MBA**....................................................................................................................... **218**

*Admissions Consulting*............................................................................................... 218

STATISTICS

# INTRODUCTION

Over the past few years, the number of GMAT (Graduate Management Admissions Test) questions involving statistics, permutations & combinations, and probability has increased substantially. Unlike the vast majority of GMAT concepts, most people do no cover these subjects in high school or college. Consequently, the makers of the test rely on these topics to create the most challenging problems, heightening the difficulty of the test. However, despite the growing number of these questions, their presence in test preparation guides and courses remains conspicuously limited. In other words, the test preparation companies are lagging behind the recent trends, short-changing their customers. *Project GMAT* addresses that need.

## A NEW FOCUS

This is the first publication that focuses entirely on statistics, permutations & combinations, and probability and offers a comprehensive coverage of all material in one source. The guide concentrates on strategies unique to the most difficult test problems and illustrates them with a number of exercises and problems devoted solely to statistics, permutations & combinations, and probability. Because it focuses on the most advanced topics, this text provides more problems and covers a wider spectrum of questions on these topics than does any other preparation guide.

## HIGH SCORES

Because of its advanced content, this text is for students aiming to score at least in the 80th percentile on the quantitative section. The reader should be familiar with the structure, content, and general strategies of the GMAT and have a good command of algebra and arithmetic. By focusing on an advanced audience, we can concentrate on the most value-added strategies and analyze the hardest questions, omitting elementary information and trivial algebraic concepts. By contrast, the topics on statistics, permutations & combinations, and probability are covered in comprehensive detail at the test level and do not assume any familiarity on the part of the reader. If you have achieved a strong command of the mathematical concepts covered on the GMAT, this guide will help you make the final push to achieve your desired score.

## BEYOND THE GMAT

Since the material on statistics and probability is a part of many quantitative courses in business school, this book will give you an advantage when you begin your MBA program. Rather than waiting to learn these subjects in school, you can master them now and gain entry into a better MBA program.

STATISTICS

# STRUCTURE OF THE BOOK

Each of the three main topics—statistics, permutations & combinations, and probability—is given its own chapter in this book. Each section starts with a thorough review of theoretical concepts and formulas, illustrated with examples and exercises. Throughout the example sections, we provide strategy tips and shortcuts aimed at improving your accuracy and time efficiency on the test. After a series of exercises, we illustrate each concept with two representative test problems, usually of intermediate or high level of difficulty, followed by explanations of the shortcuts and strategies applicable to the particular type of question. We classify each question according to its level of difficulty: from 1 (easy) to 4 (very difficult).

Each section painstakingly develops strategies unique to the particular topic—statistics, permutations & combinations, or probability. A summary of the key concepts, formulas, and strategies is included at the end of each chapter.

The last chapter of the book is devoted solely to Data Sufficiency questions, a type of problem only found on the GMAT, featuring statistics, permutations & combinations, and probability. Because a number of the distinctive strategies are applicable to advanced Data Sufficiency questions, we address this material in a separate chapter.

Finally, to respond to the growing diversity of questions at the highest levels of difficulty, this text provides a number of exercises, and 101 problems in the test format, each with detailed solutions and explanations. All of the exercises, and 36 of the problems, appear in the text to reinforce the material in the chapters. We organized the remaining 65 questions in the problem sets, by topic, providing invaluable material for practice.

STATISTICS

# STATISTICS

## STATISTICS ON THE GMAT – RECENT TRENDS

For a long time, the coverage of statistics on the GMAT was largely limited to problems involving the arithmetic average, also known as the mean. However, over the past few years, the scope of the test has been expanded to include more sophisticated questions combining the concepts of standard deviation, mode, and ranges. Despite the increase in the breadth of coverage, the <u>depth</u> of the tested theory remains quite shallow. For instance, the vast majority of the GMAT problems involving standard deviations will not ask you to compute this statistic but will simply require you to form an opinion about the relative magnitude of standard deviations. In many cases, you may be able to answer such questions even without knowing the rigorous theory – for example, just by understanding that the more spread-out sets have higher standard deviations. However, given the narrow scope of tested concepts, it is worthwhile to cover them in detail.

Primarily, in-depth knowledge of each topic will improve your level of confidence on the test and help you save additional time on stats problems. Secondly, it will prepare you for the more difficult statistics questions that may appear on future tests. Finally, since basic statistics are a part of every business school curriculum, this knowledge will serve you well during your first months at your coveted academic institution.

The purpose of this section is to familiarize you with all of the statistical concepts on the test. To demonstrate how this theory is tested, each of the topics is illustrated with exercises and sample problems. Finally, to reinforce the material, the chapter concludes with a comprehensive problem set covering the entire section.

## THE GOOD OLD MEAN

The arithmetic average remains the most popular statistical concept on the test. In actual problems, the average is often referred to as the *mean* in order to differentiate it from other averages beyond the scope of the test (such as the geometric average). For the purpose of the test, the terms *arithmetic mean*, the *mean*, and the *average* are used interchangeably and refer to the arithmetic average.

<u>**Arithmetic mean**</u> is equal to the sum of all terms divided by the number of terms.

$$\text{Arithmetic mean} = \frac{\text{sum of the terms of the set}}{\text{number of terms in the set}}.$$

## Exercise 1

The table below provides quarterly revenues of a certain company in 2002 and 2003. By what percent did the average quarterly revenue change from 2002 to 2003?

| Quarter | Quarterly Revenues, MM USD | |
|---------|------|------|
|         | **2002** | **2003** |
| 1st     | 13   | 17   |
| 2nd     | 15   | 18   |
| 3rd     | 16   | 17   |
| 4th     | 16   | 20   |

## Solution

Since we have all the quarterly revenues, we just need to calculate the averages for each of the two years and compare them.

The average quarterly revenue in 2002 $= \dfrac{13+15+16+16}{4} = \dfrac{60}{4} = 15$

The average quarterly revenue in 2003 $= \dfrac{17+18+17+20}{4} = \dfrac{72}{4} = 18$

Percent change from 2002 to 2003 $= \dfrac{18-15}{15} \cdot 100\% = 20\%$

Thus, the average quarterly revenue from 2002 to 2003 increased by 20%.

Now that we have refreshed the basic computational mechanics for the mean, we can look at more sophisticated problems that test your ability to switch gears quickly between calculating averages and calculating sums.

---

### STRATEGY TIPS

**1. On problems involving averages, be ready to move from averages to the sum of the terms and vice versa. If the problem provides an average, the solution typically requires computing the sum.**

**2. If all terms of a set are multiplied by a constant, the new average can be derived by multiplying the initial mean by the same constant.**

STATISTICS

**Problem 1**                                              **Difficulty Level: 2**

After his first semester in college, Thomas is applying for a scholarship that has a minimum Grade Point Average (GPA) requirement of 3.50. The point values of pertinent college grades are given in the table below. If Thomas took 5 courses, each with an equal weight for GPA calculations, and received two grades of A-, one grade of B+, and one grade of B, what is the lowest grade that Thomas could receive for his fifth class to qualify for the scholarship?

**Point Values of Select Grades**

| Grade | A | A- | B+ | B | B- | C+ | C | C- |
|-------|-----|-----|-----|-----|-----|-----|-----|-----|
| Value | 4.0 | 3.7 | 3.3 | 3.0 | 2.7 | 2.3 | 2.0 | 1.7 |

(A)   A

(B)   B+

(C)   B

(D)   B-

(E)   C+

## Problem 2                                                    Difficulty Level: 3

A certain portfolio consisted of 5 stocks, priced at $20, $35, $40, $45, and $70, respectively. On a given day, the price of one stock increased by 15%, while the price of another stock decreased by 35% and the prices of the remaining three remained constant. If the average price of a stock in the portfolio rose by approximately 2%, which of the following could be the prices of the shares that remained constant?

(A)   $20, $35, and $70

(B)   $20, $45, and $70

(C)   $20, $35, and $40

(D)   $35, $40, and $70

(E)   $35, $40, and $45

## SOLUTIONS: PROBLEMS 1 AND 2

### Solution to Problem 1

First, let's calculate the total number of points necessary to maintain a 3.5 average for the 5 classes. Since each class has an equal weight in the GPA calculation, the GPA can be derived by adding the points of each grade received and dividing by the number of classes taken:

$$\text{GPA} = \frac{\text{sum of points for the 5 classes}}{5}.$$

To achieve the desired GPA of 3.5, the sum of grade points for the 5 classes = $3.5 \cdot 5 = 17.5$

Since we know 4 of the 5 grades that Thomas received, we can calculate the minimum number of points that he needs to achieve the five-grade sum of 17.5:

The point sum of 4 grades = $3.7 + 3.7 + 3.3 + 3.0 = 13.7$
Minimum point value to achieve 17.5 points = $17.5 - 13.7 = 3.8$

Thus, Thomas must receive an A for the fifth class. The answer is A.

Note that by using some logic and a few shortcuts, we could have arrived at the answer even more quickly. Since A- and B+ are the same number of points away from 3.5, receiving two A- and two B+ for the four courses would yield a GPA of 3.5. Since Thomas received a B and a B+ in addition to his two A-, his current GPA must be lower than 3.5. Then to achieve a 3.5 GPA, he needs a fifth grade with a point value above 3.5, i.e. either an A or an A-. Since only A appears in the answer choices, this must be the lowest possible grade he could earn for the fifth class. Thus, the answer is A.

## Solution to Problem 2

You could solve this problem numerically by first finding the initial average, then calculating the increase in the average price of the stock and calculating the new sum of the security prices. The difference in the total value of the portfolio would give you the amount of a change in prices, which you would then use to determine the combination of securities yielding the desired result. However, this approach would be quite lengthy and prone to error. Remember, if the problem requires a series of lengthy calculations, you need to look for a shortcut.

If the overall average price of the portfolio increased, despite the larger percentage drop in one stock than the percentage increase in another, then the stock that went down in price must have been fairly cheap, while the stock that increased must have had a high price. Note that the prices of the first four securities are relatively close to each other, while the price of the fifth stock, $70, clearly stands out as a candidate to balance off the price decline. Because the percentage value of a 35% drop is more than twice the percentage value of a 15% increase, in order to get a positive ending change, the initial price of the increased stock must have been more than twice that of the reduced security. If you look at the initial prices of the stocks, the only two prices that are more than two times apart are $20 and $70. Thus, the $70 stock increased by 15%; the $20 stock declined by 35%; and the securities priced at $35, $40, and $45 did not change their prices. Therefore, the answer is E.

# MEDIAN, MODE, AND RANGE

While the arithmetic mean provides information about the average value in a set, it does not capture such summary statistics as the middle value, the most frequent element or the size of the spread of the set. This information is provided by the median, mode, and range of the set.

## 1. MEDIAN

The **median** is the middle value of a set ordered in ascending order (from least to greatest). If a set contains an even number of terms, the median is the average of the two middle values.

---
### STRATEGY TIPS
---

**When calculating the median, always rearrange the terms in ascending or descending order before finding the middle value. If the set contains an even number of terms, average the two middle values.**

---

Let's consider an example.

## Exercise 2

Set A consists of numbers {-2, 27.5, -6, 18.3, 9} and set B consists of values {-199, 0.355, 19.98, 10, 201, 16}. The median of set B is how much greater than the median of set A?

## Solution

First, let's arrange the terms of both sets in ascending order: set A {-6, -2, 9, 18.3, 27.5} and set B {-199, 0.355, 10, 16, 19.98, 201}. Because set A contains an odd number of terms, its median is simply the middle value, 9. Set B contains an even number of terms, so its median is the average of the two middle values, 10 and 16: $\frac{10+16}{2} = 13$. The difference between the medians of the two sets is $13 - 9 = 4$.

## 2. MODE

The **mode** is the most frequently occurring value in a set. If several values occur with the same frequency, each of them represents a mode, and the sequence has several modes. Note that if each value in a set occurs just once, then it does not mean that the set has no mode; in this case, each of the values in the set is a mode.

## Exercise 3

Sets A, B, and C consist of the following elements:

Set A {0, 3, 4, 2, 0, 4, 7, 8, 4, 17}
Set B {20, 12, -7, -9, -5, -7, 11, -5, 68}
Set C {-1.5, 0, 1.5}

If function Z is defined as the sum of the modes of sets A, B, and C, what is the value of Z?

## Solution

In set A, number 4 appears most often – three times. Therefore, the mode of set A is 4. In set B, the most frequently appearing values are -5 and -7, each of which appears twice. Thus, the modes of set B are -5 and -7. In Set C, each of the three values appears once, so set C has 3 modes: -1.5, 0, and 1.5.

Now, we can compute the sum of the modes of the three sets: $Z = 4 - 5 - 7 - 1.5 + 0 + 1.5 = -8$.

### 3. RANGE

The **range** of a set is the difference between the largest and the smallest value in a set. To find the range of the set, subtract the smallest value from the largest. There are several properties of the range that you should remember. First, the range of the set is always non-negative. Second, if the range of the set is zero, all numbers in the set must be identical. Finally, changing any values other than the smallest and the largest has no effect on the range.

## Exercise 4

If Set X contains numbers {-21, 6, 19, 126, 1000} and Set Y contains values {-21, 990, 993, 996.19, 997.05, 999, 1000}, what is the difference between the ranges of Set X and Set Y?

## Solution

The range of set X = 1000 – (-21) = 1021; the range of set Y = 1000 – (-21) = 1021. Difference between the two ranges = 1021 – 1021 = 0.

The previous example illustrates the fact that sets with the same smallest and largest values will always have the same ranges, regardless of the number of terms in each set or the distribution of elements between the largest and the smallest value.

---

**STRATEGY TIPS**

---

Range of a set is determined only by the largest and smallest values and does not depend on the number of terms in a set or their distribution between the largest and smallest values. Consequently, adding or dropping elements between the largest and the smallest values has no effect on the range.

---

## Problem 3                                                    **Difficulty Level: 3**

Set X consists of prime numbers $\{3, 11, 7, K, 17, 19\}$. If integer Y represents the product of all elements in set X and if 11Y is an even number, what is the range of set X?

(A)  14

(B)  16

(C)  17

(D)  20

(E)  26

## Problem 4                                    Difficulty Level: 4

Set A consists of integers {3, -8, Y, 19, -6,} and Set B consists of integers {K, -3, 0, 16, -5, 9}. Number L represents the median of Set A, number M represents the mode of set B, and number $Z = L^M$. If Y is an integer greater than 21, for what value of K will Z be a divisor of 26?

(A)  -2

(B)  -1

(C)  0

(D)  1

(E)  2

## SOLUTIONS: PROBLEMS 3 AND 4

### Solution to Problem 3

Since 11Y is an even integer, Y must be even. Because Y is equal to the product of all elements in set X, at least one of the numbers in set X must be even. Since all elements in set X are primes, and 2 is the only even prime number, K = 2.

The range of Set X = 19 – 2 = 17. The answer is C.

Note also that 17 is the only odd number among the answer choices. Since we know that all the numbers in set X are odd, except for K, the only way to receive an odd range of the set is to have an even number as its largest or smallest value. Because all numbers in set X are primes, K = 2 (the only even prime) and the range = 19 – 2 = 17.

### Solution to Problem 4

First, since we know that Y > 21, we can arrange the values of Set A in ascending order: {-8, -6, 3, 19, Y}. We can now see that regardless of the actual value of Y, the median of Set A = 3. Thus, L = 3, and $Z = 3^M$, implying that Z will be equal to some power of 3.

However, note that because 26 is not divisible by 3, the only integer power of 3 that will yield a factor of 26 is 0, yielding $Z = 3^0 = 1$. Therefore, we know that M = 0 and that the mode of Set B is equal to 0.

Finally, for Set B to have a single mode of 0, this value has to appear most frequently (i.e. twice) among the 5 terms of the set. It follows that K = 0. The answer is C.

# STANDARD DEVIATION

As we have seen before, the range of a set does not fully capture the degree of its dispersion, providing information only about the two extreme values in a set. In real life, it is often important to summarize the consistency in the data and to analyze its spread around the mean—the concept captured by the standard deviation. Knowing how closely the set is distributed around its mean is often important, for instance, in evaluating the accuracy of the forecasts. Let's consider an example that illustrates the meaning of standard deviation.

## Example

Suppose you are the manager of a factory and need to purchase the raw materials for the next month in advance. The amount of raw materials that you need depends on the demand for your products, which varies from month to month. You estimate that the average demand is likely to be 15,000 units. Thus, you know the mean of the set. However, because you have no further information, you do not know by how much your demand will fluctuate around its mean; nor do you know the likelihood of any specific demands.

If your demand is fairly consistent, with possible values of 14,000, 14,500, 15,000, 15,500 and 16,000, you will have much more confidence in the average value of 15,000 and will wish to hold a limited amount of safety stock. Consider now that your demand can take values 5,000, 10,000, 15,000, 20,000 and 25,000. Clearly, the second scenario holds significantly more uncertainty about the amount of needed materials and will require a substantial level of safety stock. In the second case, the average value cannot be used as an accurate estimate of the future demand. Although the two sets have the same mean, the difference in their variability is captured by the concept of standard deviation.

**Standard deviation** assesses how closely the terms in the set are spread around its mean. This is done by calculating an average difference between each term in the set and the mean. However, because some terms in the set are greater than the mean and some are smaller than the mean, the differences between the terms and the mean can be both positive and negative. If these differences were simply averaged, then positive deviations would be netted against negative, thus artificially reducing the actual size of the spreads.

To ensure that the absolute values of positive and negative differences are added (rather than netted against each other), these differences are squared and then added to find the total sum of squared deviations in a set. This sum is then divided by the number of terms to get the average square deviation. Finally, a root is taken from the average square deviation in order to take away the increase in the magnitude caused by the initial squaring.

STATISTICS

---

## STRATEGY TIPS

---

**To find a standard deviation of a set, complete the following steps:**

1. **Find the mean**

2. **Compute the differences between the mean and each number in a set**

3. **Square these differences and add them together**

4. **Divide the sum of the squared differences by the number of terms[1]**

5. **Take the square root of the result**

---

It is unlikely that you will be asked to calculate the value of a standard deviation on the test. However, you should be able to analyze and compare the magnitudes of standard deviations of different sets, i.e. recognize which sets have smaller, larger, or equal standard deviations. You should also be able to tell what will happen to the standard deviation in a set if the set undergoes some transformation (for example, if all numbers in a set are multiplied by a constant).

---

[1] In some cases in statistics, the sum of the squared differences is divided by the number of terms less 1, i.e. by N–1; for the purpose of the test, you can always divide by the number of terms, i.e. by N.

Let's consider a simple example that involves calculating the standard deviation.

## Exercise 5

Jack played four bowling frames. He knocked down 8 pins in his first frame, 9 pins in his second frame, 7 pins in his third frame and 10 pins in his last frame. What is the standard deviation of the number of pins knocked down in the four frames?

## Solution

First, let's find the average number of pins that were knocked down:

$$\text{Mean} = \frac{8+9+7+10}{4} = \frac{34}{4} = 8.5$$

Now, let's compute the differences between the number of pins knocked down in each frame and the average and square these differences:

| Frame | Result | Mean | Result minus mean | Differences squared |
|-------|--------|------|-------------------|---------------------|
| 1st | 8 | | $8 - 8.5 = -0.5$ | $(-0.5)^2 = 0.25$ |
| 2nd | 9 | 8.5 | $9 - 8.5 = 0.5$ | $0.5^2 = 0.25$ |
| 3rd | 7 | | $7 - 8.5 = -1.5$ | $(-1.5)^2 = 2.25$ |
| 4th | 10 | | $10 - 8.5 = 1.5$ | $1.5^2 = 2.25$ |

Find the sum of the squared differences and divide it by the number of terms:

$$\text{Average squared difference} = \frac{0.25 + 0.25 + 2.25 + 2.25}{4} = 1.25.$$

Finally, we can find the standard deviation by taking the square root of the average squared difference:

$$\text{Standard deviation} = (1.25)^{\frac{1}{2}} \approx 1.12.$$

Note that the result of each of the four attempts (8, 9, 7, and 10 pins) was either 0.5 or 1.5 away from the mean of 8.5. Thus, the value of 1.12 lies approximately in between these deviations and captures the average deviation of attempts from the mean. Note however, that because of the squaring, the standard deviation is not equivalent to an average absolute difference (which would be equal to 1 in this case: $\frac{0.5 + 0.5 + 1.5 + 1.5}{4} = 1$).

Now we are ready to try some practice test problems.

STATISTICS

**Problem 5**                                                    **Difficulty Level: 3**

Set A consists of all prime numbers between 10 and 25; Set B consists of consecutive even integers, and Set C consists of consecutive multiples of 7. If all the three sets have an equal number of terms, which of the following represents the ranking of these sets in an ascending order of the standard deviation?

(A)   C, A, B

(B)   A, B, C

(C)   C, B, A

(D)   B, C, A

(E)   B, A, C

**Problem 6**                                     **Difficulty Level: 3**

Set A consists of all even integers between 2 and 100, inclusive. Set X is derived by reducing each term in set A by 50, set Y is derived by multiplying each term in set A by 1.5, and set Z is derived by dividing each term in set A by -4. Which of the following represents the ranking of the three sets in descending order of standard deviation?

(A)   X, Y, Z

(B)   X, Z, Y

(C)   Y, Z, X

(D)   Y, X, Z

(E)   Z, Y, X

## SOLUTIONS: PROBLEMS 5 AND 6

### Solution to Problem 5

First, let's find the terms of set A by listing all prime numbers between 10 and 25:

Set A {11, 13, 17, 19, 23}.

Because the three sets have an equal number of terms, there must be 5 elements in set B and in set C. Since the members of sets B and C can be any consecutive multiples of 2 and 7, respectively, let's pick small numbers to illustrate the trend:

Set B {0, 2, 4, 6, 8};   Set C {7, 14, 21, 28, 35}.

Just by a quick look, we can see that the spread of the terms is the greatest in set C. Therefore, Set C has the highest standard deviation. By comparing sets A and B, we can conclude that the terms of set B increase at the step of 2, while the elements of the prime set increase at steps ranging from 2 to 4. Also, note that the five elements in Set A cover the range of 12, while the five elements in Set B cover the range of only 8. As a result, we can conclude that the elements of Set B are more closely grouped and that Set B has a lower standard deviation than set A. Therefore, the correct ranking of the three sets in an increasing order of their standard deviations is B, A, C. The answer is E.

Note that it did not matter which specific five consecutive multiples of 2 and 7 we selected. Although the values of the numbers would be different, the rate of increase in the numbers and the difference between each element and the mean would remain the same, resulting in the same standard deviation. For instance, if we picked Set C to represent {1007, 1014, 1021, 1028, and 1035}, the differences between the mean (1021) and each of the elements would still be the same as in the set {7, 14, 21, 28, 35} with a mean of 21. These differences would be 14, 7, 0, -7, and -14. Hence, the two sets must have the same standard deviation.

---

### STRATEGY TIPS

**Adding or subtracting a constant from each element in the set has no effect on the standard deviation.**

---

STATISTICS

## Solution to Problem 6

As we learned in the last problem, adding or subtracting a constant from each element in a sequence does not affect the differences between the terms, as all of them are shifted simultaneously on a number line. Thus, the standard deviation of Set X will be equal to that of Set A.

For Set Y, multiplying each element of a set by a constant with an absolute value greater than or equal to 1 will increase the differences between the terms, spreading the set over a larger range. For instance, by multiplying the first and the last terms of Set A by 1.5, you can see that this set will contain numbers between 3 and 150. Intuitively, since the number of terms will not change but the range will increase, the spread between the numbers will increase and so will the standard deviation.

Finally, dividing each term of a set by -4 is equivalent to multiplying by -0.25. Clearly, the signs of all elements in the set will be reversed. More importantly, the differences between the terms will shrink by 4 times. Note that the same number of elements in Set Z will now cover a range between -0.5 and -25 (range = 24.5), compared with the initial range of $100 - 2 = 80$. Thus, this division will make elements in Set Z closer to one another, reducing the standard deviation.

In summary, the standard deviation of Set X will remain the same, that of Set Y will increase, and that of Set Z will decrease, compared with the initial set A. Therefore, the correct ranking of standard deviations from the highest to the lowest (note that the problem asks for descending order) is Y, X, Z. The answer is D.

---

### STRATEGY TIPS

**When finding a standard deviation of a set:**

1. **Multiplying each term of a set by a number with an absolute value greater than 1 increases the standard deviation, while multiplying by a number with an absolute value less than 1 decreases the standard deviation.**

2. **Dividing each element in a set by a number with an absolute value greater than 1 decreases the standard deviation, while dividing by a number with an absolute value less than 1 increases the standard deviation.**

3. **Changing the sign of all elements in the set or multiplying by (-1) has no effect on the standard deviation.**

STATISTICS

# THE MAP STRATEGY FOR STANDARD DEVIATION

This strategy will help you remember what happens to the standard deviation of a set after various transformations. In the sample maps, pay attention to the distance between the towns.

### MULTIPLYING AND DIVIDING – CHANGE THE SCALE

You can think of multiplying or dividing all numbers in the set as the same process as changing the scale of a map. If you increase the scale of the map by multiplying by an absolute value greater than 1, each distance on the map will increase proportionally. The average distance from the mean (standard deviation) will also increase. Similarly, if you decrease the scale of the map, the distances will shrink, resulting in a decrease in the standard deviation.

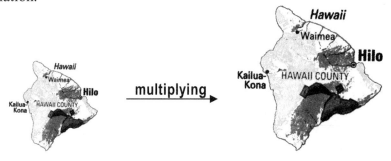

### CHANGING SIGNS – MIRROR IMAGE

To remember that the standard deviation is not affected by a simultaneous shift in the sign of each element, compare the switching of sign to the process of making an exact mirror image of the map. If you make a 1:1 mirror reflection of the distances on the map, their magnitude will not change, keeping the standard deviation constant.

### ADDING OR SUBTRACTING CONSTANTS – SHIFT THE MAP

If a constant is added to or subtracted from each term in the set, this action is similar to shifting the entire map to the left on the number line (in case of an increase) or to the right (in case of a decrease). The relative values of your distances will not be affected by moving an entire map. Therefore, the standard deviation of the set will not change.

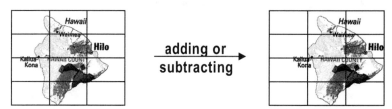

STATISTICS

# CHAPTER SUMMARY

## 1. CONCEPTS AND FORMULAS

**<u>Arithmetic mean</u> (or simply the mean)** – the average value in a set.

$$\text{Arithmetic Mean} = \frac{\text{sum of the terms of the set}}{\text{number of terms in the set}} .$$

---

**<u>Median</u>** – the middle value of a set ordered in ascending order. If a set contains an even number of terms, the median is the average of the two middle values.

**To find the median**, always rearrange the terms in ascending or descending order before finding the middle value. If a set contains an even number of terms, average the two middle values.

---

**<u>Mode</u>** – the most frequently occurring value in a set. If several values occur with the same frequency, each of these values represents a mode, i.e. the sequence has several modes.

**To find the mode**, identify the most frequently occurring value(s) in a set.

---

**<u>Range</u>** – the difference between the largest and the smallest value in a set.

**To find the range,** subtract the smallest value from the largest.

---

**<u>Standard deviation</u>** measures how closely the terms in the set are spread around its mean.

**To find a standard deviation:**
1. Find the mean
2. Compute the differences between the mean and each number in a set
3. Square these differences and add them together
4. Divide the sum of the squared differences by the number of terms
5. Take the square root of the result

---

STATISTICS

## 2. STRATEGIES AND TIPS

- On problems involving averages, be ready to calculate both the mean and the sum. If the problem provides an average, the solution typically requires computing the sum.

- If all terms of a set are multiplied by a constant, the new average can be derived by multiplying the initial mean by the same constant.

- If the smallest and the largest value of the set do not change, manipulations with all other values have no effect on the range

- Use the Map Strategy to deduce the effect on standard deviation:

    o Adding or subtracting a constant from each element in the set has no effect on the standard deviation. Shifting a map does not change the distances.

    o Multiplying each term of a set by a number with an absolute value greater than 1 increases the standard deviation, while multiplying by a number with an absolute value less than 1 decreases the standard deviation. Increasing the scale of the map increases distances, while reducing the scale shrinks them.

    o Dividing each element in a set by a number with an absolute value greater than 1 decreases the standard deviation, while diving by a number with an absolute value less than 1 increases the standard deviation. Increasing the scale of the map increases distances, while reducing the scale shrinks them.

    o Changing the sign of all elements in the set or multiplying by (-1) has no effect on the standard deviation. Taking a 1:1 mirror image of a map has no effect on the magnitude of distances.

STATISTICS

# STATISTICS PROBLEM SET

**10 QUESTIONS, 20 MINUTES**

**Problem 7**                                              **Difficulty level: 2**

If John makes a contribution to a charity fund at school, the average contribution size will increase by 50%, reaching $75 per person. If there were 5 other contributions made before John's, what is the size of his donation?

(A)   $100

(B)   $150

(C)   $200

(D)   $250

(E)   $450

## Problem 8

**Difficulty level: 2**

What is the minimum percentage increase in the mean of set X {-4, -1, 0, 6, 9,} if its two smallest elements are replaced with two different primes?

(A)   25%

(B)   50%

(C)   75%

(D)   100%

(E)   200%

**Problem 9**                                      **Difficulty level: 2**

If M is a negative integer and K is a positive integer, which of the following could be the standard deviation of a set {-7, -5, -3, M, 0, 1, 3, K, 7}?

I.  -1.5
II. -2
III. 0

(A)    I only

(B)    II only

(C)    III only

(D)    I and III only

(E)    None

**Problem 10**                                    **Difficulty level: 2**

The median annual household income in a certain community of 21 households is $50,000. If the mean income of a household increases by 10% per year over the next 2 years, what will the median income in the community be in 2 years?

(A)   $50,000

(B)   $60,000

(C)   $60,500

(D)   $65,000

(E)   It cannot be determined from the information given

**Problem 11**                                    **Difficulty level: 3**

Sets A, B and C are shown below. If number 100 is included in each of these sets, which of the following represents the correct ordering of the sets in terms of the absolute increase in their standard deviation, from largest to smallest?

A {30, 50, 70, 90, 110}
B {-20, -10, 0, 10, 20}
C {30, 35, 40, 45, 50}

(A)   A, C, B

(B)   A, B, C

(C)   C, A, B

(D)   B, A, C

(E)   B, C, A

**Problem 12**                                    **Difficulty level: 3**

Which of the following series of numbers, if added to the set {1, 6, 11, 16, 21}, will not change the set's mean?

I.   1.50, 7.11 and 16.89
II.  5.36, 10.70, 13.24
III. -21.52, 23.30 and 31.22

(A)   I only

(B)   II only

(C)   III only

(D)   II and III only

(E)   None

**Problem 13**                                          **Difficulty level: 3**

Set X consists of all two-digit primes and set Y consists of all positive odd multiples of 5 less than 100. If the two sets are combined into one, what will be the range of the new set?

(A)   84

(B)   89

(C)   90

(D)   92

(E)   95

**Problem 14**                                          **Difficulty level: 3**

What is the range of a set consisting of the first 100 multiples of 7 that are greater than 70?

(A)   693

(B)   700

(C)   707

(D)   777

(E)   847

## Problem 15

**Difficulty level: 3**

Which of the following could be the median of a set consisting of 6 different primes?

(A)   2

(B)   3

(C)   9.5

(D)   12.5

(E)   39

**Problem 16**                                                    **Difficulty level: 4**

If numbers N and K are added to set X {2, 8, 10, 12}, its mean will increase by 25%. What is the value of $N^2 + 2NK + K^2$?

(A)   28

(B)   32

(C)   64

(D)   784

(E)   3600

## SOLUTIONS: STATISTICS PROBLEM SET

### Answer Key:

7. C    10. E    13. D    16. D
8. D    11. E    14. A
9. E    12. C    15. E

### Solution to Problem 7

Since the new average contribution of $75 is 1.5 times greater than the initial contribution, we know that the initial average was $50.

Let's denote the amount of John's contribution as J and the sum of other five contributions as T and use the formula for the mean to construct the following equations:

$$\text{Old average} = \frac{T}{5} = 50; \quad T = 250$$

$$\text{New Average} = \frac{T+J}{6} = \frac{250+J}{6} = 75$$
$$250 + J = 450 \implies J = 200$$

The answer is C.

### Solution to Problem 8

Since all prime numbers are positive, replacing -4 and -1 (the two smallest elements) with two primes will always increase the sum of the terms of set X and consequently will raise its mean. To achieve the smallest increase in the mean, we need to replace the two negative numbers with the smallest primes. Since both primes have to be different, these numbers will be 2 and 3. Thus, after the replacement, set X will contain elements {0, 2, 3 6, 9}. Using this information, we can find the new mean and the percentage increase in the mean:

$$\text{Mean of Set X before replacement} = \frac{-4-1+0+6+9}{5} = \frac{10}{5} = 2$$

$$\text{Mean of Set X after replacement} = \frac{0+2+3+6+9}{5} = \frac{20}{5} = 4$$

$$\text{Percentage increase in the mean} = \frac{4-2}{2} \cdot 100\% = 100\%$$

The answer is D.

## Solution to Problem 9

The standard deviation measures the absolute average spread between the mean and other elements. Since the differences between the mean and other elements are squared and added, the value of the standard deviation will always be non-negative, eliminating answers A, B and D.

Further, because the differences between the mean and each element are squared, positive and negative differences accumulate in the absolute value of the standard deviation rather than cancel each other. Thus, the only way to have a 0 standard deviation is to have all elements in the set equal to each other and equal to the mean. Since we know that all elements cannot be equal regardless of the values of K and M, the standard deviation of our set cannot be 0, eliminating answer D. Therefore, none of the given values can represent the standard deviation of the set. The answer is E.

## Solution to Problem 10

Since there are 21 households in the community, the median income will be equal to the income of the household that occupies the $11^{th}$ place in the ranking of incomes. From the information about the changes in the mean income, we can find the new total of all incomes in the community. However, we have no information about how that income will be allocated among the households.

For example, if incomes of all families increased proportionally by 10% per year, the new median would be $60,500 (i.e. $50,000 \cdot 1.1^2$). On the other hand, if the entire increase in the average income were attributable solely to the richest household, the median income would remain at $50,000. Therefore, there is insufficient information to answer the question. The answer is E.

## Solution to Problem 11

Just by inspection, we can see that the 3 sets are evenly spaced, contain the same number of elements and have the means of 70, 0 and 40, respectively. Since the standard deviation measures the average distance between each element and the mean, the absolute increase in the standard deviation will be proportional to how far the new number will be from the mean. In other words, the greater the absolute value of the difference between the current mean and the new number, the greater the distortion that will be caused by the new number and the greater the increase in the standard deviation.

Since the mean of set A (70) is closest to 100, the increase in the standard deviation of set A will be the smallest. Similarly, the increase in the standard deviation of set B (mean = 0) will be the largest and the increase in the standard deviation of set C (mean = 40) will be the second largest. Thus, the correct ranking from largest to smallest increase in standard deviation is B, C, A. The answer is E.

## Solution to Problem 12

If the problem involves numbers that are cumbersome to manipulate, look for shortcuts. If we look at the initial set, we can see that each number in the set is 5 greater than the previous number. Because the elements are evenly spaced, the middle element will be the mean of the set. Thus, without any computational work, we know that the mean of the original set is 11.

In order to preserve the mean of the original set, the 3 added numbers must also have the mean of 11. Therefore, the sum of the 3 new numbers must be 33.

Now, we can quickly go through the answer choices, adding only integer parts to see if we can get a number close to 33. Adding the integer parts of the first and second sets will give us numbers that are too low – 24 and 28, respectively. Adding the integer values of the third set will yield: -21 + 23 + 31 = 33. Thus, set 3 is worth a closer look.

If we look at the decimal parts in the third set, we can see that they will add up to 0: -0.52 + 0.30 + 0.22 = 0. Thus, the sum of the elements in the third set is equal to 33 and its average is equal to 11. Consequently, adding the elements of the third set to the original set will have no effect on the mean. The answer is C.

## Solution to Problem 13

To find the range of the new set, we need to determine its smallest and largest values. Since the smallest element in set X is 11 and the smallest element in set Y is 5, the smallest element in the combined set will be 5.

The largest element in set Y is 95. However, we cannot determine whether this will be the largest value in the combined set without knowing the largest element in set X. If this element is greater than 95, then the greatest value in the combined set will be equal to this element. If the largest element in set X is less than or equal to 95, then the largest value in the combined set will be 95. Thus, we need to find whether any member of set X is greater than 95.

Since set X includes only two-digit primes, we must check whether there are any prime numbers among 96, 97, 98 and 99. We can immediately exclude even numbers 96 and 98. We also know that 99 is divisible by 3 and hence cannot be a prime number. Thus, we need to check only 97.

Since 97 is not divisible by 2, 3, 5 or 7, it is a prime. Note that we need to check the divisibility of 97 only by the prime factors that are less than its square root (approximately 10). If 97 were divisible by any primes greater than 10, the quotient would always be less than 10 and we have already checked for all factors less than 10 and would have identified them if they existed. Thus, 97 is a prime number and hence belongs to set X. Therefore, the largest value in set X and in the combined set is 97.

Range of the combined set = 97 – 5 = 92. The answer is D.

---

## STRATEGY TIPS

**To check whether a number is a prime, check its divisibility only by the factors that are less than the square root of the number.**

---

## Solution to Problem 14

To find the range of the set, we need to determine both its first (smallest) member and its last (largest) element. The first multiple of 7 greater than 70 is $7 \cdot 11 = 77$. Thus, the first and smallest member in the set will be 77. To understand the trend in the sequence, you can see that the second term will be $7 \cdot 12 = 84$, the third term will be $7 \cdot 13$, the fourth term will be $7 \cdot 14$ and so on.

As you can see, each next term in the set is derived by increasing the multiplier of 7 by 1. Since there are 99 more elements after the first term of 77, in order to find the $100^{th}$ multiple of 7, we need to increase the multiplier of 7 from the first term by 99.

Multiplier of 7 for the $100^{th}$ = $11 + 99 = 110$

$100^{th}$ term = $7 \cdot 110 = 770$

Range of the set = $770 - 77 = 693$

The answer is A.

## Solution to Problem 15

Since the set contains 6 terms, its median will be equal to the average of the third and fourth term, if the elements are arranged in ascending order. Since the set contains different primes, 2 (the smallest prime) and 3 (the second smallest prime), cannot be the third or the fourth term, nor can they be the median of the set, eliminating choices A and B.

Since all primes other than 2 are odd, the sum of the third and fourth elements will always be even. Therefore, when we divide this sum by 2 to find the average, we will always get an integer (which could be even or odd), eliminating answers C and D.

Thus, we are left with choice E, the correct answer. Indeed, 39 would be the median of the set if the third and fourth largest elements were 37 and 41, respectively.

## Solution to Problem 16

Since $N^2 + 2NK + K^2 = (N + K)^2$, to answer the question in the problem, we need to find N+K. First, let's calculate the mean of set X and then use this value to find the sum of the terms in the new set.

Sum of the terms of set X = 2 + 8 + 10 + 12 = 32

Original mean of set X = $\dfrac{\text{sum of the terms}}{\text{number of terms}} = \dfrac{32}{4} = 8$

Mean of set X after numbers N and K are added = $8 \cdot 1.25 = 10$

Sum of the terms of set X after numbers N and K are added = Mean · Number of terms = $6 \cdot 10 = 60$

Now we can find the sum of N and K as the difference between the sum of the terms of set X before and after numbers N and K are added.

N + K = 60 – 32 = 28

$(N + K)^2 = 28^2 = 784$

The answer is D.

# COMBINATORICS

## COMBINATORICS MADE EASY

The term *combinatorics* refers to an area of mathematics that deals with the arrangements of different items, or permutations & combinations. Some examples of combinatorial problems include estimating the number of different ways to arrange five cards on a table, to sit ten people at a corporate reception, or to select four elective courses at business school.

Since combinatorics are not covered in a typical high school or college curriculum, these problems are usually classified as upper-bin material. Test-takers aiming at the 90th quantitative percentile will have to correctly solve several problems dealing with combinatorics. If that was not enough, the test makers often incorporate combinatorial concepts into problems on other topics, especially in probability theory. Despite the increasing prevalence of the combinatorics problems, some prep courses still recommend solving them by inspection, i.e. by writing out all possible combinations and counting them. While this intellectually rigorous approach may yield the desired result, it is very time consuming, prone to error, and difficult to apply to problems with a large number of possibilities and restrictions.

To build the necessary toolkit for solving combinatorics problems quickly and efficiently, this section begins with coverage of the essential theoretical concepts followed by examples and strategies for solving each type of combinatorics problem. Finally, the section concludes with a comprehensive set of problems covering all combinatorial topics.

## WHEN DOES THE ORDER MATTER?

Combinatorics problems can be divided into two broad categories – those for which the order of arrangements is important (permutations) and those for which the order does not matter (combinations). To provide some insight into these concepts, let's consider a few examples demonstrating the differences between ordered and unordered arrangements.

### PERMUTATIONS

*Permutations,* or scenarios where the sequence of elements is important, include the arrangement of letters in a company name, digits on a credit card, or students in a classroom. Each variation in the sequence (e.g. switching places of letters, digits, or students) will generate a different permutation.

Another distinguishing feature of permutations lies in the fact that the number of elements often corresponds to the number of available slots for these elements. Typical problems involving permutations include arranging people in a line (every person needs one slot) or sitting an audience in a movie theatre (each person must be seated).

## COMBINATIONS

By contrast, questions dealing with *combinations* (unordered arrangements) refer to arrangements in which the order does not matter and where a smaller group of elements is drawn from a larger pool. Examples of such problems include selecting a three-person case competition team from a class of 250 MBA students, choosing several items from a menu at a fast-food restaurant, or purchasing three suits at a department store to wear to your second-round interviews. In each of these cases, the sequence of choices is unimportant and does not create a new combination.

For instance, if students John, Jenny, and Lisa are selected to represent the school at a case competition, changing the order in which they were chosen does not yield a new combination. For instance, a team consisting of Lisa, John, and Jenny selected in that order is the same as the original team where John was selected first, followed by Jenny and Lisa. Similarly, if you order a turkey sandwich, then salad, then a glass of orange juice, you will end up with the same meal, regardless of which item you ordered first.

If the order is not important in combinations, then what is? New combinations are created by selecting a different set of items from the pool of available possibilities. For instance, in our menu example, ordering the same turkey sandwich and a salad with apple juice rather than orange juice would constitute a new combination of a meal. Similarly, replacing any (or all) of the three students with other candidates on the case competition team would also result in a new combination.

# PERMUTATIONS: THE WORLD IN ORDER

As we discussed earlier, permutations represent different orderings (sequences) of elements. Let's illustrate this concept with a simple example.

### Exercise 6

Along with your handsome job offer, you have been invited on an all-expenses-paid trip to Europe, a courtesy of the employer desperately trying to recruit you before graduation. On your trip, you would like to see London, Paris, Rome, and Madrid, which can be visited in any sequence. How many different travel itineraries, defined as arrangements of the four destination stops, are possible?

### Solution

Let's try to visualize this problem by using a simple diagram. In the scheme below, the four cells represent the number of choices for the first, second, third, and fourth city visited on your trip.

| 1$^{st}$ city | 2$^{nd}$ city | 3$^{rd}$ city | 4$^{th}$ city |
| --- | --- | --- | --- |
|  |  |  |  |

When we start making our travel plans, we can first visit any of the four cities, a property reflected by the four available possibilities for filling in the first slot. Once the first city to be visited has been selected – let's suppose it is London – we can choose one of the three remaining cities (Paris, Madrid or Rome) for our second stop, an outcome reflected by three different possibilities in the cell for the second city. Similarly, after two of the first destinations have been decided, we have only two possibilities for selecting the third stop. Finally, after the first three destinations have been determined, we are left with only one choice for the fourth stop. The diagram below shows the number of possibilities available for each of the four stops on the European trip.

| 1$^{st}$ city | 2$^{nd}$ city | 3$^{rd}$ city | 4$^{th}$ city |
| --- | --- | --- | --- |
| 4 | 3 | 2 | 1 |

Thus, for each of the four combinations for the first stop, we have three possibilities for the second stop, for each of which we again have 2 possibilities for the third stop and one for the final stop. Then the total number of different itineraries can be derived by multiplying the number of possibilities for each of the four stops on the European trip. Thus, we can compute the total number of different itineraries for your trip:

Number of itineraries = $4 \cdot 3 \cdot 2 \cdot 1 = 24$.

# FACTORIALS!

As we saw in the previous example, the number of different routes that can be created by rearranging the four city stops can be computed by multiplying all the consecutive integers starting from 1 and ending with 4 – the number of city stops that we can make. Thus, the more cities we can visit, the larger the number of different routes that we can create.

For instance, if we had a chance to visit 5 cities instead of 4, then the total number of different itineraries would be equal to $1 \cdot 2 \cdot 3 \cdot 4 \cdot 5 = 120$. If we could visit 6 cities, then the number of different routes in which we could arrange then on our trip would equal $6 \cdot 5 \cdot 4 \cdot 3 \cdot 2 \cdot 1 = 720$.

Let's look more closely at this concept. In general, when we need to evaluate the number of permutations that can be created by a rearrangement of N distinct elements, we need to compute the number of ways in which these elements can be allocated among the N different spots that we have for them (one spot per one element). Then, as we saw in the example with the European trip, we can choose any one of the N elements to be placed in the first spot (N alternatives for the first spot). After we fill the first spot, we will have used up one of the N elements and are left with N–1 alternatives for the second spot. Similarly, we will have N–2 candidates for the third spot, N–3 elements for the fourth spot, and so on, until we get to the last spot. Since the number of spots equals the number of elements, by the time we get to the last spot, there will be just one element left to fill that spot. Thus, we will have just one choice for the last spot, as illustrated in the diagram below:

| N | N – 1 | N – 2 | N – 3 | ... | 1 |
|---|-------|-------|-------|-----|---|

Then the total number of different arrangements will be equal to the product of the number of choices that we have for each of the spots:

The total number of arrangements = $N \cdot (N–1) \cdot (N–2) \cdot (N–3) \cdot \ldots \cdot 1$.

Note that this expression is a product of all consecutive integers from 1 to N, where N is the number of elements to be rearranged. Since this product is used very frequently in combinatorics, it received a special name – *factorial*. Factorial of any number is denoted by an exclamation sign following the number; for instance, 5! denotes the product of $1 \cdot 2 \cdot 3 \cdot 4 \cdot 5$ and is read "five factorial."

The table below shows some examples of factorials:

| Factorial | Pronounced | Value |
|-----------|------------|-------|
| 3! | Three factorial | $1 \cdot 2 \cdot 3 = 6$ |
| 7! | Seven factorial | $1 \cdot 2 \cdot 3 \cdot 4 \cdot 5 \cdot 6 \cdot 7 = 5,040$ |
| K! | K factorial | $1 \cdot 2 \cdot \ldots \cdot (K–1) \cdot K$ |

> **Factorial of a number N (denoted N!) is a product of all positive consecutive integers up to N, inclusive:  N! = 1 · 2 ·...· (N–1) · N**
>
> **The number of ways in which N distinct items can be arranged = N!**

Since factorial denotes the product of positive integers smaller than or equal to N, negative factorials do not exist. Note that factorial of zero is defined to be equal to 1, i.e. 0! = 1. While this is somewhat counter-intuitive, defining 0! = 1, as opposed to 0, is done to avoid division by 0 in combinatorics problems.

Although you can always compute the value of a factorial by writing out the product and multiplying the integers, it is worthwhile to memorize the following factorials that appear frequently on the test:

### Factorials to remember

| Factorial | Value |
|:---:|:---:|
| 0! | 1 |
| 1! | 1 |
| 2! | 2 |
| 3! | 6 |
| 4! | 24 |
| 5! | 120 |
| 6! | 720 |

## Exercise 7

Tanya bought five glasses for her kitchen in five different colors – white, red, black, grey, and blue – and would like to display them on the same shelf next to each other. In how many different ways can Tanya arrange the glasses?

## Solution

Since Tanya must display all the glasses on the shelf, the number of arrangements of the glasses can be computed as a factorial of the number of distinct items – factorial of five. Thus, the number of arrangements = 5! = 5 · 4 · 3 · 2 · 1 = 120.

Now, let's demonstrate how factorials can be applied to increase your time efficiency on the test problems.

> **CIRCULAR**
>
> <u>A common specific problem:</u>  How many different arrangements of x elements are there if they are all arranged in a circle?
>
> $$(x - 1)!$$

COMBINATORICS

## Problem 17               Difficulty Level: 2

Sergey, a second-year MBA student, applied for the position of associate at each of the 8 investment banks recruiting at his school. After returning from his fall break, he finds out that he received an invitation for an interview from 50% of the firms to which he applied and decides to accept all interview invitations. If each firm where he received an invitation is conducting one first-round interview, which Sergey can schedule at his discretion, what is the number of ways in which he can design his interview schedule, defined as the sequence of first-round interviews?

(A)    4

(B)    8

(C)    24

(D)    32

(E)    64

**Problem 18**                                                    **Difficulty Level: 3**

Janice is decorating her house by arranging Christmas lights along the top of her entrance door. The Christmas lights come in 7 different colors and each of the lights costs 39 cents at regular price. Janice has 2 dollars to spend on Christmas lights, and she buys the lights during a holiday sale when the store gives away 1 light for every 4 purchased at the regular price. What is the maximum number of light decorations she can create if she has to use all the lights purchased?

(A)   5

(B)   6

(C)   24

(D)   120

(E)   720

## SOLUTIONS: PROBLEMS 17 AND 18

### Solution to Problem 17

Since Sergey received an invitation from half of the 8 firms, he will need to schedule 4 first-round interviews. Because each interview schedule will consist of the 4 first-round interviews, the number of different schedules that can be constructed will equal the number of ways to arrange the 4 interviews in different order. Therefore, the number of possible interview schedules = $4! = 4 \cdot 3 \cdot 2 \cdot 1 = 24$. The answer is C.

### Solution to Problem 18

By quick inspection, you should see that 2 dollars is not evenly divisible by 39 cents. Rather than dividing 200 cents by 39 cents, you can save time by approximating the price of the light at 40 cents, since the additional one cent on each light will be insufficient to buy an additional item and thus will not change the number of lights purchased. Thus, at the regular price, Janice would be able to buy $\dfrac{2\,\text{dollars}}{40\,\text{cents}} = 5$ lights.

However, since she bought the lights during the holiday sale and received an additional light for each 4 purchased, she was able to buy 6 lights. Since she would like to create the maximum number of arrangements, she must have purchased the lights of different color. As a result, Janice has 6 lights of different color to use in her door decoration.

Since changing the ordering of the colors creates a new ornament, the order does matter and our task is to find the number of different permutations of 6 items. Because all of the purchased lights must be used in decoration, the total number of different ornaments is equal to the number of ways in which the 6 lights can be placed in 6 spots, i.e. the number distinct ordered arrangements (permutations) of 6 elements. Thus, the total number of decorations = $6! = 720$. The answer is E.

---

### STRATEGY TIPS

1. **Challenging test problems typically do not require extensive computational work. When a GMAT problem gives a number that is difficult to manipulate, look for ways to approximate the value without changing the result of the operation. If you find yourself performing extensive computations, look for a shortcut and check your work to make sure that you did not make an arithmetic error.**

2. **To distinguish between combinations and permutations, try to change the order of the items. If a change in order yields a new arrangement, apply the formulas for permutations; otherwise, use the formulas for combinations.**

COMBINATORICS

# PERMUTATIONS WITH REPEATING ELEMENTS

In all of our earlier examples, we assumed that each of the elements was distinct – every Christmas light was of different color, every investment banking interview was with a specific firm, and every stop on European trip had to be made in a different city. While these interpretations are common in intermediate problems, questions of higher difficulty often involve repeating elements. Examples of these problems include computing the number of arrangements of 5 pencils, two of which are identical, or finding the number of different words that can be created with a set of letters, where some letters are the same.

> The number of arrangements of N items, K of which are the same $= \dfrac{N!}{K!}$

Intuitively, since some items are the same, the number of possible different arrangements is reduced. If all of the N items were distinct, the number of different orderings would be N! – the expression that appears at the top of the formula. However, since K of the N items are the same, the opportunities to create new orderings are reduced by the number of permutations that could be created from the K items – K!, which appears at the bottom of the fraction.

Let's demonstrate the applications of this formula.

## Exercise 8

Thomas bought 2 black, 1 red, and 1 green pencil, all of which are identical, except for color. If he aligns all the pencils along the side of a table and places them parallel to each other with the sharpened end pointing in the same direction, how many different arrangements can he create?

## Solution

Since Thomas must place all 4 pencils on the table, the only way to create new arrangements is to change the order of the pencils. Thus, the order matters and we will use the formula for permutations. Further, because 2 of the black pencils are the same, we will be arranging 4 items (N = 4), of which 2 are identical (K = 2). Using the formula, we can find that the total number of arrangements $= \dfrac{4!}{2!} = \dfrac{24}{2} = 12$.

If there were more than one group of similar items, i.e. if Thomas had 2 black and 2 red pencils, then the number of permutations would be equal to $\dfrac{4!}{2! \cdot 2!}$.

> If K of the N elements are of the same type and L of the N elements are also of the same type, the number of distinct arrangements of the N elements equals $\dfrac{N!}{K! \cdot L!}$

COMBINATORICS

Now, let's modify our example with the Christmas lights by limiting the number of available colors to illustrate the effect of repeating elements:

## Problem 19                                            Difficulty Level: 3

Janice is decorating her house by arranging Christmas lights along the top of her entrance door. The Christmas lights come in 5 different colors and each of the lights costs 39 cents at regular price. Janice has 2 dollars to spend on Christmas lights, and she buys the lights during a holiday sale when the store gives away 1 light for every 4 purchased at the regular price. What is the maximum number of light decorations that she can create if she must use all the lights purchased?

(A)   5

(B)   60

(C)   120

(D)   360

(E)   720

**Problem 20**                                        **Difficulty Level: 3**

Cliff is contemplating a name for a new sailboat that he recently purchased. He has a set of copper letters from the word PEPPER left from the name of his earlier sailboat, and plans to use all of them for his new boat. If he can use only the letters from the previous boat, how many different names can he create, assuming that a name of the boat does not necessarily have to be a meaningful word?

(A)   6

(B)   60

(C)   120

(D)   360

(E)   720

## SOLUTIONS: PROBLEMS 19 AND 20

### Solution to Problem 19

In Problem 18, we found that Janice will be able to purchase 6 lights. Because she would like to maximize the number of different decorations, she will buy as many distinct colors as possible. Since there are only 5 available colors, Janice will buy 5 lights of distinct colors and 1 light that will be identical to one of those that she has already purchased. Thus, the number of possible decorations will equal the number of arrangements of 6 items, two of which are identical. By using the formula for permutations with repeating elements for N = 6 and K = 2, we can find the total number of possible arrangements:

$$\frac{N!}{K!} = \frac{6!}{2!} = \frac{720}{2} = 360$$

Therefore, the answer is D.

### Solution to Problem 20

Since Cliff must use all of the letters of the word PEPPER and cannot use any other letters, he can create new names only by rearranging the 6 available letters, i.e. N = 6. Also, note that the word PEPPER has three identical letters P and two letters E, i.e. K = 3 and L = 2, factors that will reduce the number of distinct names. By using the formula for permutations with repeating elements, we can find the total number of distinct names:

$$\frac{N!}{K! \cdot L!} = \frac{6!}{3! \cdot 2!} = \frac{1 \cdot 2 \cdot 3 \cdot 4 \cdot 5 \cdot 6}{(1 \cdot 2 \cdot 3) \cdot (1 \cdot 2)} = 2 \cdot 5 \cdot 6 = 60$$

Thus, the answer is B.

### STRATEGY TIPS

**When performing computations with factorials, write out the products of integers for all factorials and reduce similar terms before performing other operations.**

COMBINATORICS

## PERMUTATIONS INVOLVING SELECTION

So far, we have only considered problems where the number of elements in the final set was equal to the total number of elements in the pool from which they were drawn. For instance, all the four cities had to be visited on the European trip, all Christmas lights had to be used in a holiday decoration, and all letters of the word PEPPER had to appear in the new name of a new sailboat. In other words, the new permutations could be created only by rearranging the elements, without dropping or adding any new items.

In many GMAT problems, you will be asked to calculate the number of possible arrangements that could be created by selecting a set of elements from a larger pool. For instance, the earlier problems would fall into this category if we were allowed to visit just 3 of the 4 European cities, to use only 4 of the 6 lights for the holiday decoration, and to create new sailboat names from any 5 letters of the word PEPPER.

In these types of problems, new arrangements can be obtained not only by rearranging the selection of elements but also by selecting other elements from the initial pool. For example, suppose that we can visit only 3 of the 4 cities on the European trip, and we initially decide to visit London, Paris, and Madrid in that order. We can create new itineraries in one of the two ways:

(1) rearranging the 3 selected cities (e.g. changing London-Paris-Madrid to Madrid-London-Paris)
(2) replacing one of the cities in our selection (e.g. replacing London with Rome to create a new route Rome-Paris-Madrid).

---

**In the problems where the order of elements is relevant and where the final selection is smaller than the initial pool, i.e. K elements have to be selected from a pool of N items, the number of different permutations is calculated by a formula:**

$$\text{Total number of permutations} = \frac{N!}{(N-K)!}$$

---

N = the number of elements in the pool
K = the number of elements that must be selected from the pool

We will refer to this formula as the **Permutations Formula**, since it can be applied to the vast majority of problems dealing with <u>ordered</u> arrangements. Note that the earlier class of problems – in which all of the N elements of the pool had to be in the final selection of K items – represents just a special case of this formula. If all available elements have to be used, then N = K and N – K = 0, yielding the number of permutations = $\frac{N!}{0!} = \frac{N!}{1} = N!$

### STRATEGY TIPS

**If you wish to limit the amount of memorization for permutation problems, memorize the general permutations formula and adjust it for the special cases.**

Let's modify some of our earlier examples to illustrate this formula.

## Exercise 9

Tanya bought 5 glasses for her kitchen – a white, red, black, grey, and blue – and would like to display 3 of them on one shelf next to each other. In how many different ways can Tanya arrange the glasses?

## Solution

Since all the glasses are of different color, displaying them in different order will create new arrangements. Therefore, because the order is relevant, we are in the permutations world and we can use the permutations formula to calculate the total number of different ways in which Tanya can select 3 from the 5 glasses and rearrange them on the shelf.

The total number of ways to display the glasses $= \dfrac{N!}{(N-K)!} = \dfrac{5!}{(5-3)!} = \dfrac{5!}{2!} = 60$.

Thus, Tanya can create 60 different arrangements of the three glasses.

Some GMAT problems do not specify the number of elements that must be selected but rather define the lower or upper bound on the final selection, i.e. "select at least K elements from N" or "at most K elements from N." In this case, the total number of permutations can be computed by adding the number of ways to select each allowable number of elements. For instance, the number of ways to select at least 4 elements from a pool of 6 will be equal to the sum of ways to select 4, 5, and 6 items. Similarly, the number of permutations that can be created by selecting at most 3 elements from a pool of 7 is equal to the sum of ways to select 1, 2, and 3 elements.

---

### STRATEGY TIPS

The total number of ways to select <u>at least</u> or <u>at most</u> K elements can be found by computing the number of ways to select each allowable number of elements, i.e. ways to select K, K + 1, …, N elements for "at least" problems and 1, 2, …, K elements for "at most" problems.

---

Let's modify our earlier examples to illustrate this concept.

## Exercise 10

Anna has to visit at least 2 European cities on her vacation trip. If she can visit only London, Paris, Rome, or Madrid, how many different itineraries, defined as the sequence of visited cities, can Anna create?

## Solution

Since Anna has to visit at least 2 of the 4 capitals, she can stop in 2, 3, or 4 of the cities. Because an itinerary is defined as an ordered sequence, i.e. a route from Rome-Madrid is not the same as a route from Madrid-Rome, the order matters and we are in the land of permutations.

The total number of different itineraries that can be created will be equal to the sum of all two-city, three-city, and four-city routes, computed from the permutations formula $\frac{N!}{(N-K)!}$ with N = 4 and K = 2, 3, or 4.

Number of two-city itineraries $= \frac{4!}{(4-2)!} = \frac{4!}{2!} = 12$

Number of three-city itineraries $= \frac{4!}{(4-3)!} = \frac{4!}{1!} = 24$

Number of four-city itineraries $= \frac{4!}{(4-4)!} = \frac{4!}{0!} = 24$

Total number of possible routes = 12 + 24 + 24 = 60

Thus, if Anna has to visit at least 2 of the 4 European capitals, she can create 60 different itineraries.

COMBINATORICS

Let's illustrate how these concepts are applied to some of the more difficult test problems.

## Problem 21                                              Difficulty level: 3

Mike, a DJ at a high-school radio station, needs to play two or three more songs before the end of the school dance. If each composition must be selected from a list of the 10 most popular songs of the year, how many song sequences are available for the remainder of the dance?

(A)    6

(B)    90

(C)    120

(D)    720

(E)    810

COMBINATORICS

### Problem 22                              Difficulty level: 4

A password to a certain database consists of digits that cannot be repeated. If the password is known to consist of at least 8 digits and it takes 12 seconds to try one combination, what is the amount of time, in minutes, necessary to guarantee access to the database?

(A)    $\dfrac{8!}{5}$

(B)    $\dfrac{8!}{2}$

(C)    $8!$

(D)    $\dfrac{10!}{2}$

(E)    $\dfrac{5}{2} \cdot 10!$

## SOLUTIONS: PROBLEMS 21 AND 22

### Solution to Problem 21

Because Mike can play either two or three songs before the end of the dance, the total number of music selections will equal to the number of two- and three-song sequences that can be created from the list of the top 10 singles. Because the sequence in which the compositions are played is important, we are dealing with permutations. Thus, we can compute the number of two- or three-element permutations created from the pool of 10 elements using the permutations formula for N = 10 and K = 2 or 3:

Number of two-song selections: $\dfrac{10!}{(10-2)!} = \dfrac{10!}{8!} = 9 \cdot 10 = 90$

Number of three-song selections: $\dfrac{10!}{(10-3)!} = \dfrac{10!}{7!} = 8 \cdot 9 \cdot 10 = 720$

Total number of ways to finish the dance = 90 + 720 = 810

Thus, Mike has 810 choices to finish the dance. Therefore, the answer is E.

### Solution to Problem 22

Since there are 10 distinct digits (0 through 9), none of which can be repeated, the password can contain at most 10 digits. Given that there are at least 8 digits in the password, it can contain 8, 9, or 10 digits. Because rearranging the digits leads to a new password, the order of elements is relevant and we can apply the permutations formula.

The number of permutations of 8, 9, and 10 elements from the pool of 10 can be computed from the permutations formula $\dfrac{N!}{(N-K)!}$ , where N = 10 and K = 8, 9, or 10.

Number of eight-digit passwords = $\dfrac{10!}{(10-8)!} = \dfrac{10!}{2!} = \dfrac{10!}{2}$

Number of nine-digit passwords = $\dfrac{10!}{(10-9)!} = \dfrac{10!}{1!} = 10!$

Number of ten-digit passwords = $\dfrac{10!}{(10-10)!} = \dfrac{10!}{0!} = 10!$

Total number of possible passwords = $\dfrac{10!}{2} + 10! + 10! = 10!\,(0.5 + 1 + 1) = \dfrac{5}{2} \cdot 10!$

Since it takes 12 seconds ($\frac{1}{5}$ of a minute) to try one password, we can calculate the time necessary to try all combinations:

Time to try all passwords = $\dfrac{5}{2} \cdot 10! \cdot \dfrac{1}{5} = \dfrac{10!}{2}$ minutes. Therefore, the answer is D.

# COMBINATIONS: IN THE WORLD OF ANARCHY

In the previous sections we considered situations when the order of elements was important – changing the order of items resulted in a new arrangement. In another large class of problems, the order of items is irrelevant. Some of these situations include selecting a committee of 5 members from 10 nominees, choosing 3 office locations from 7 major cities, or allocating 4 job offers among 10 interviewees. In each of these cases, mere rearrangement of the elements (i.e. a change in the sequence in which the committee members are selected or job offers extended) does not create a new combination. Therefore, the order in which the elements are arranged is unimportant.

For instance, if job offers were extended to Lisa, Mark, Lena, and Terry in that sequence, changing the order in which the job offers were extended, (e.g. to Mark, Lisa, Terry, and Lena) will not result in a new group of hires. The only way to create a new combination would be to add, drop, or replace one or more of the selected elements. For example, if 4 offers have to be extended, a new combination of recruits can be created by extending an offer to, say, Frank rather than Mark (presuming that the offers will be accepted, of course).

Because rearranging the elements in questions where the order is irrelevant will not create a new combination, all combination problems deal with selecting a smaller subset from a larger pool. Otherwise, if all elements from the pool have to be included in the selection and the order of items is unimportant, there is only one way to make this choice. For instance, there is only one way to select a group of 5 new hires from 5 interviewees or to choose a team of 8 participants from 8 nominees. No additional combinations can be created.

---

**The number of unordered arrangements consisting of K items selected from a pool of N elements is computed from the Combinations Formula:**

$$\text{Number of unordered arrangements} = \frac{N!}{K! \cdot (N - K)!}$$

---

Note that this expression is quite similar to the permutations formula. The only difference is that K! appears in the denominator of the combinations formula and thus reduces the resulting number of total arrangements. The intuition behind this difference is that the number of ordered arrangements created from a pool of elements has to be greater than or equal to the number of unordered arrangements. If a mere rearrangement of the items yields a new variation, we can create far more new arrangements than in the case when the order of items is irrelevant. Therefore, for a given size of the pool and number of elements in the selection, the number of possible permutations (ordered arrangements) will always be greater than or equal to the number of combinations (unordered arrangements).

## STRATEGY TIPS

To determine which formula to use for permutations and which for combinations, remember that the number of combinations is smaller than the number of permutations and that the denominator of the combinations formula has to be greater, i.e. it must include K!

Permutations ≥ Combinations

$$\frac{N!}{(N-K)!} \geq \frac{N!}{K! \cdot (N-K)!}$$

Let's consider a few problems that address combinations.

# Exercise 11

Jose needs to select 3 people for his study group in a core finance class. If there are 8 other students in the class still seeking a group, in how many ways can he form his group?

# Solution

Since the order in which Jose selects his teammates will not affect the composition of his team, we are dealing with combinations. Thus, we need to calculate the number of unordered arrangements of 3 elements (3 spots remaining on Jose's team) that can be created from a pool of 8 students. To compute this, we can use the combinations formula with N = 8 and K = 3:

Number of ways to fill the team $= \dfrac{N!}{K! \cdot (N-K)!} = \dfrac{8!}{3! \cdot 5!} = 56$

Thus, Jose can create his group in 56 ways.

## SIMPLE COMBINATIONS

If you choose one object from a set of *m* objects and another from *n* objects, there are *m* x *n* different possibilities of combining the two objects.

For example, say you have a jar with 10 balls numbered from 1 through 10 and a jar with 5 balls marked with the letters A – E, if you pick one ball from each jar there are 10 x 5 = 50 number-letter combinations.

**Problem 23**                                                    **Difficulty Level: 3**

The government of country Forellia, the location of an international soccer tournament, needs to allocate 4 identical broadcasting licenses among the TV stations that have submitted the applications and meet the established criteria. The country has 20 TV stations, three-quarters of which applied for the broadcast license. If 40% of the applicants do not meet the established criteria, how many different allocations are possible?

(A)   45

(B)   60

(C)   126

(D)   512

(E)   1024

**Problem 24**                                    **Difficulty Level: 4**

A university needs to select a nine-member committee on extracurricular life, whose members must belong either to the student government or to the student advisory board. If the student government consists of 10 members, the student advisory board consists of 8 members, and 6 students hold membership in both organizations, how many different committees are possible?

(A)   72

(B)   110

(C)   220

(D)   720

(E)   1096

## SOLUTIONS: PROBLEMS 23 AND 24

### Solution to Problem 23

Since three-quarters of the 20 stations applied for a license, the government must have received 15 applications. Because 40% of the applicants do not meet the established criteria, the number of qualified candidates will be reduced to 9. Because the government will award 4 licenses, we need to estimate the number of 4-element sets (licensees) that can be created from a pool of 9 elements. Since all the licenses are identical, the order in which they are awarded is not relevant. Thus, we need to use the combinations formula for N = 9 and K = 4 to compute the number of allocations:

Number of possible allocations =

$$= \frac{N!}{K! \cdot (N-K)!} = \frac{9!}{4! \cdot 5!} = \frac{1 \cdot 2 \cdot 3 \cdot 4 \cdot 5 \cdot 6 \cdot 7 \cdot 8 \cdot 9}{(1 \cdot 2 \cdot 3 \cdot 4) \cdot (1 \cdot 2 \cdot 3 \cdot 4 \cdot 5)} = \frac{5 \cdot 6 \cdot 7 \cdot 8 \cdot 9}{1 \cdot 2 \cdot 3 \cdot 4 \cdot 5} = \frac{7 \cdot 8 \cdot 9}{4} = 7 \cdot 2 \cdot 9 = 126$$

Therefore, the answer is C.

Note how we simplified the expression with three factorials by writing out the products and canceling like terms. Use this technique to save time and improve your accuracy on the test.

### Solution to Problem 24

To compute the number of possible committees, we need to find the size of the pool from which we need to select the 9 members of the committee, i.e. to determine the number of people who are members of student government, advisory board, or both. Since 6 students hold membership in both of these organizations, 6 of the 8 members in the advisory board are included in the 10 members of the student government, while 2 of the 8 students are not counted in the student government.

The total number of students who hold membership in at least one of the two organizations is equal to the sum of members of the student government (10 people) and members of the advisory board who do not belong to the student government (2 people).

To compute the number of the 9-member committees that can be selected from a pool of 12 candidates, we can use the combinations formula with N = 12 and K = 9:

Number of committees = $\dfrac{12!}{9! \cdot (12-9)!} = 220$

Therefore, the answer is C.

# ADVANCED TOPICS IN COMBINATORICS

The topics we have considered so far cover the vast majority of the combinatorics problems on the GMAT. However, for the readers aiming at the highest scores on the quantitative section, this chapter offers additional material focused on the most difficult combinatorics problems.

One of the most popular ways to increase the difficulty of a combinatorics problem is to add additional restrictions and constraints to an intermediate or already difficult question. For example, recall the question about the European trip. To raise the complexity of the problem, the test authors may specify that if the student visits London, she must also visit Paris and cannot visit both Madrid *and* London. Similarly, in the question about glasses on a kitchen shelf, additional constraints may require that the black glass not appear next to the blue glass and that the white or grey glass stand at the end of the sequence. Because the constraints for each problem are unique, there is no hard and fast rule that can be applied uniformly to all of the constraint problems. However, in the majority of cases, the most efficient approach is to start with an unconstrained scenario and then make adjustments for the constraints. Let's modify some of our earlier examples to illustrate this approach.

## Exercise 11

Tanya bought 5 glasses for her kitchen – white, red, black, grey, and blue – and would like to display 3 of them on the shelf next to each other. If she decides that a black and a white glass cannot be displayed at the same time, in how many different ways can Tanya arrange the glasses?

## Solution

### Approach 1

The first approach to solving a constrained problem is to compute the number of arrangements ignoring the constraint and then exclude the arrangements that do not meet the specified conditions. The base-case number of arrangements (no constraints) can be computed by using the permutations formula for an ordered 3-element selection from a pool of 5 elements:

$$\text{Number of base-case arrangements} = \frac{5!}{(5-3)!} = \frac{5!}{2!} = 60$$

Now we need to exclude the situations when both the white and black glass appear among the three selected for display. To compute the number of arrangements to be excluded, we need to find the number of ways in which we can select both a white and a black glass from the pool of 5 glasses bought by Tanya. Since the order in which the glasses are placed in this problem is important, we will use the permutations formula to compute the number of exclusions. If both the white and the black glass are selected, then we have 3 of the 5 glasses left (blue, grey, and red) to fill the one remaining slot on the shelf (2 other spots are already occupied by the white and black glass). Further, note that there are three choices for the remaining slot – it can be on the left, in the middle, or on

the right. Finally, for each location of the remaining slots, there are two ways that the black and white glasses can be arranged (black followed by white or white followed by black). As a result, the number of exclusions can be computed as the product of the number of possibilities to fill the remaining one slot (3 glasses), the number of locations of the remaining slot (3 locations), and the number of ways in which the black and white glass can be arranged with respect to each other for every location of the remaining slot (2 ways).

Number of exclusions = $3 \cdot 3 \cdot 2 = 18$

Thus, the number of allowable arrangements = $60 - 18 = 42$

### Approach 2

The second approach to solving constrained problems is to consider from the very start only the combinations meeting the constraint. If a white and black glass cannot be displayed together, then we have 3 possible scenarios:

(1) Only the white glass is displayed.
(2) Only the black glass is displayed.
(3) Neither of the two glasses is displayed.

We can then compute the total number of possible arrangements by calculating the number of possible arrangements for each of the three scenarios and then summing up the results.

If only the white glass is displayed, it can appear in any of the 3 spots. Additionally, we will have 3 choices left to fill the three remaining slots. Note that the black glass will be automatically excluded from the pool of 5 and the white glass has already been used up, yielding $5 - 1 - 1 = 3$ choices for the remaining two slots. Thus, to fill the remaining two slots, we need to compute the number of arrangements to choose a 2-element ordered set from the pool of three elements:

Scenario (1): Number of arrangements for two other glasses $\dfrac{3!}{(3-2)!} = 3! = 6$

Since the white glass can occupy any spot among the three displayed glasses (on the left, in the middle, or on the right), we will have $3 \cdot 6 = 18$ possibilities for Scenario 1.

Scenario (2) will be identical to Scenario (1), with the only exception that a black glass rather than white will be displayed. Thus, the number of arrangements for the second scenario is also equal to 18.

Scenario (3) presumes that neither the black nor white glass is displayed. Thus, we are left with 3 choices of glasses (red, grey, and blue) to fill the 3 slots on the shelf. Since the number of elements in the pool is equal to the number of spots in the final set, the number of variations will be equal to the number of rearrangements that can be done among the 3 elements, i.e. $3! = 6$.

The total number of possible arrangements will be equal to $18 + 18 + 6 = 42$.

---

## STRATEGY TIPS

---

**To solve constrained combinatorics problems, use one of the two approaches:**

1. **Solve the problem without the constraint and exclude the cases that do not satisfy the constraint.**

   **OR**

2. **Consider only the scenarios that meet the constraints and sum up the total number of possibilities for each of the scenarios.**

---

**Problem 25**                                    **Difficulty Level: 4**

An engagement team consists of a project manager, team leader, and four consultants. There are 2 candidates for the position of project manager, 3 candidates for the position of team leader, and 7 candidates for the 4 consultant slots. If 2 of the 7 consultants refuse to be on the same team, how many different teams are possible?

(A)   25

(B)   35

(C)   150

(D)   210

(E)   300

**Problem 26**                                          **Difficulty Level: 4**

A nickel, a dime, and two identical quarters are arranged along a side of a table. If the quarters and the dime have to face heads up, while the nickel can face either heads up or tails up, how many different arrangements of coins are possible?

(A)   12

(B)   24

(C)   48

(D)   72

(E)   96

## SOLUTIONS: PROBLEMS 25 AND 26

### Solution to Problem 25

First, let's find the total number of ways to fill the 4 consulting slots, ignoring the constraint. Since the 4 positions are equivalent and the order in which the consultants are selected is unimportant, we can calculate this number from the combinations formula with N = 7 and K = 4:

$$\text{Number of ways to select consultants (no constraint)} = \frac{7!}{4! \cdot (7-4)!} = 35$$

Second, let's calculate the number of combinations that violate the constraint and have to be excluded from the total count. If both of the "enemy" consultants are chosen, then there are 5 candidates left to fill the remaining 2 slots on the team. Thus, the number of exclusions will be equal to the number of ways to select the 2 remaining consultants from the pool of 5 people:

$$\text{Number of exclusions} = \frac{5!}{2! \cdot (5-2)!} = 10$$

The number of ways to select consultants for the project = 35 −10 = 25

For each of the 25 possible groups of the consultants, there are 2 possibilities for an engagement manager and 3 possibilities for a team leader. Therefore, the total number of engagement teams that can be created can be computed as a product of the possibilities for each of the three components of the team:

Total number of engagement teams = 25 · 2 · 3 = 150. The answer is C.

### Solution to Problem 26

First, let's calculate the number of possible arrangements of coins, ignoring the specifications about heads and tails. Because the coins of different denominations represent distinct elements, the order is important and we are dealing with permutations. Further, note that all of the four elements in the pool will be included in the final selection and that two of the elements (two quarters) are identical. Therefore, we need to apply the formula for permutations with repeating elements. Recall that the number of N-element selections with K repeating elements is equal to $\frac{N!}{K!}$.

$$\text{Number of coin arrangements, ignoring heads/tails} = \frac{4!}{2!} = 12$$

Because the nickel can appear either heads or tails, we need to expand the number of arrangements. Since the nickel is included in each of the coin arrangements, allowing the

COMBINATORICS

coin to show up either heads or tails will double the total number of possible arrangements. Intuitively, we calculated that there are 12 ways to arrange all coins showing up heads. Now, we can create one twin for each of the 12 arrangements by turning the nickel so that it shows up tails:

Total number of arrangements = 12 · 2 = 24. The answer is B.

# STRATEGIES UNIQUE TO COMBINATORICS

## 1. PERMUTATIONS OR COMBINATIONS? SEQUENCE IT!

To memorize that unordered arrangements refer to combinations and that ordered arrangements yield permutations, note that the sequence of letters "na" in the word combiNAtions is repeated in aNArchy.

## 2. SKETCH IT!

Many of the test problems on combinatorics involve a limited number of possible arrangements. If the number of elements is small (e.g. you are asked to find the number of different ways to sit four people or to assign three offices), you can often draw out the possible arrangements and count them. Even if you cannot come up with all the different possibilities, this approach will help you eliminate some or all incorrect answer choices. Finally, drawing a picture will often help you visualize the problem, actively analyze it, and draw parallels with similar questions you have seen before, providing you with additional clues to the right solution.

## 3. COMBINATORICS DECISION TREE

This strategy will provide a framework that you can apply to the majority of combinatorics problems and will help you structure your approach, determine the appropriate formulas, and apply sequential analysis eliminating paths at every step. In many cases, it is difficult to get started on a challenging combinatorics problem. This framework will provide you with a way to begin your analysis and narrow down the number of approaches to the problem. To deconstruct a test problem, apply a series of standard questions:

**1. Does changing the order of elements create a new arrangement?**

If yes, the order is important and we are dealing with permutations; proceed with path A
If no, the order is irrelevant and we are dealing with combinations; proceed with path B

**Path A (permutations)**

**A1. Are any of the elements identical?**

If yes, use the formula for permutations with repeating elements.

If no, use the general permutations formula.

**A2. Are there any additional constraints in the problem?**

If yes, adjust your results in step 2 to account for the scenarios that do not satisfy the constraints and get the final answer.

If no, check whether you have considered all possible cases and select the final answer.

COMBINATORICS

**Path B (combinations)**

**B1. Are there any additional constraints in the problem?**

If yes, calculate the number of combinations from the combinations formula and adjust your results according to the constraints; then select the final answer.

If no, apply the combinations formula, check whether you have considered all scenarios, and select the final answer.

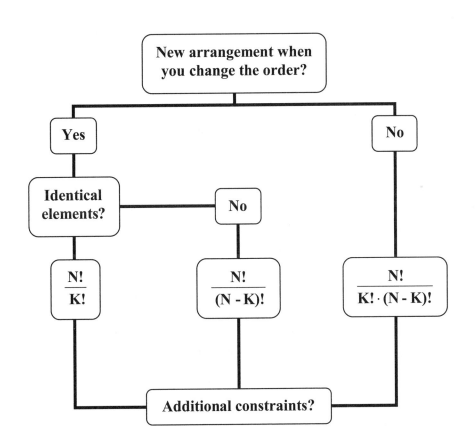

# CHAPTER SUMMARY

## 1. CONCEPTS AND FORMULAS

**Factorial** – a product of all positive consecutive integers less than or equal to the number under the factorial.

Examples:

$$N! = 1 \cdot 2 \cdot \ldots \cdot (N{-}1) \cdot N \qquad\qquad 9! = 1 \cdot 2 \cdot 3 \cdot 4 \cdot 5 \cdot 6 \cdot 7 \cdot 8 \cdot 9 = 362,880$$

Factorials to remember:

| Factorial | Value |
|-----------|-------|
| 0! | 1 |
| 1! | 1 |
| 2! | 2 |
| 3! | 6 |
| 4! | 24 |
| 5! | 120 |
| 6! | 720 |

**Circular:** If x elements are all arranged in a circle, the number of different arrangements are: **$(x - 1)!$**

---

**Permutations** – ordered arrangements of elements

Examples:
 (1) Letters in a password
 (2) Colors in a color coding

**Number of ways to arrange N distinct items = N!**

**Permutations with repeating elements** – permutations of elements, some of which are indistinguishable from others.

Examples:
 (1) Arrangements of different coins, two of which are identical quarters
 (2) Arrangements of letters, some of which are repeated

**Number of arrangements of N items, K of which are the same $= \dfrac{N!}{K!}$**

**General Permutations Formula:**

**Number of ordered K-element arrangements created from N elements $= \dfrac{N!}{(N - K)!}$**

**Combinations** – unordered arrangements of elements.

Examples: members or a committee or students in a study group

**General Combinations Formula:**

**Number of unordered K-element sets created from N elements** = $\dfrac{N!}{K! \cdot (N-K)!}$

**Simple Combinations:** If you choose one object from a set of *m* objects and another from *n* objects, there are *m x n* different possibilities of combining the two objects.

---

## 2. STRATEGIES AND TIPS

- To distinguish between combinations and permutations, try to change the order of the items. If this change yields a new arrangement, apply the formulas for permutations; otherwise, use the formulas for combinations.
- To remember that unordered arrangements refer to combinations and that ordered arrangements yield permutations, note that the sequence of letters "na" in the word combiNAtions is repeated in aNArchy.
- Use drawings on problems involving a limited number of elements.
- To structure your approach to combinatorics problems, use the Combinatorics Decision Tree
- When performing computations with factorials, write out the products of integers and reduce similar terms before performing other operations.
- To distinguish between the general permutations & combinations formulas, remember that the number of combinations is smaller than the number of permutations and that the denominator of the combinations formula has to be greater, i.e. it must include K!
- The number of ways to select <u>at least</u> or <u>at most</u> K elements can be found by computing the number of ways to select each allowable number of elements, i.e. ways to select K, K + 1, …, N elements for "at least problems" and 1, 2, …, K elements for "at most" problems.
- To solve constrained combinatorics problems, use one of two approaches:

  (1) Solve the problem without the constraint and exclude the cases that do not satisfy the constraint.
  OR
  (2) Consider only the scenarios that meet the constraints and sum up the total number of possibilities for each of the scenarios.

# COMBINATORICS PROBLEM SET

**10 QUESTIONS, 20 MINUTES**

**Problem 27**                                    **Difficulty level: 1**

Tanya is taking a final examination that contains 5 essay questions, of which she must answer 3. How many different final exams could Tanya potentially create?

(A)    10

(B)    15

(C)    20

(D)    60

(E)    120

## Problem 28                                    **Difficulty level: 1**

Assuming that there are no ties, how many different outcomes are possible in a race with 5 athletes?

(A)  25

(B)  120

(C)  240

(D)  360

(E)  720

**Problem 29**                                                   **Difficulty level: 1**

Melissa has 6 favorite books. If she plans to take 3 books on her transatlantic flight, how many choices does she have to make her selection?

(A)   6

(B)   18

(C)   20

(D)   24

(E)   120

**Problem 30**                                    **Difficulty level: 2**

At a certain laboratory, chemical substances are identified by an unordered combination of 3 colors. If no chemical may be assigned the same 3 colors, what is the maximum number of substances that can be identified using 7 colors?

(A)   21

(B)   35

(C)   105

(D)   135

(E)   210

**Problem 31**                                    **Difficulty level: 2**

An equity analyst needs to select 3 stocks for the upcoming year and rank these securities in terms of their investment potential. If the analyst has narrowed down the list of potential stocks to 7, in how many ways can she choose and rank her top 3 picks?

(A)   21

(B)   35

(C)   210

(D)   420

(E)   840

## Problem 32

**Difficulty level: 2**

How many different five-letter combinations can be created from the word TWIST?

(A)   5

(B)   24

(C)   60

(D)   120

(E)   720

**Problem 33**  **Difficulty level: 3**

If an employee ID code must consist of 3 non-repeating digits and each digit in the code must be a prime number, how many ID codes can be created?

(A)   12

(B)   24

(C)   60

(D)   120

(E)   240

## Problem 34                                    **Difficulty level: 3**

A university cafeteria offers 4 flavors of pizza – pepperoni, chicken, Hawaiian and vegetarian. If a customer has an option (but not the obligation) to add, extra cheese, mushrooms, or both to any kind of pizza, how many different pizza varieties are available?

(A)   4

(B)   8

(C)   12

(D)   16

(E)   32

**Problem 35**                                        **Difficulty level: 4**

A book store has received 8 different books, of which $\frac{3}{8}$ are novels, 25% are study guides and the remaining are textbooks. If all books must be placed on one shelf displaying new items and if books in the same category have to be shelved next to each other, how many different arrangements of books are possible?

(A)    18

(B)    36

(C)    72

(D)    216

(E)    432

## Problem 36

**Difficulty level: 4**

A group of 5 students bought movie tickets in one row next to each other. If Bob and Lisa are in this group, what is the probability that both of them will sit next to only one other student from the group?

(A)  5%

(B)  10%

(C)  15%

(D)  20%

(E)  25%

## SOLUTIONS: COMBINATORICS PROBLEM SET

### Answer Key:

27. A  30. B  33. B  36. B
28. B  31. C  34. D
29. C  32. C  35. E

### Solution to Problem 27

Because changing the order of the questions does not create a new exam, we are in the world of combinations. To answer the question, we need to compute the number of 3-element unordered selections that can be created from a pool of 5 elements.

$$\text{Number of 3-question exams} = \frac{5!}{3! \cdot (5-3)!} = 10$$

The answer is A.

### Solution to Problem 28

Since there are no ties in a running competition, the number of outcomes is equal to the number of ways to assign the 5 athletes to places from first to fifth. In other words, the total number of outcomes is equal to the number of ways to rearrange the 5 runners from best to worst. The number of rearrangements that can be created from 5 elements is equal to 5! or 120. The answer is B.

### Solution to Problem 29

Since switching the order of selected books does not create a new selection, we need to use the combinations formula:

$$\text{Number of possible selections} = \frac{6!}{3! \cdot (6-3)!} = 20. \text{ The answer is C.}$$

### Solution to Problem 30

Since no chemical may be assigned the same 3 colors as another, a mere rearrangement of these colors will not enable us to mark another chemical. Therefore, the order does not matter and we are dealing with combinations. We can find the number of 3-element unordered selections from the pool of 8 elements by using the combinations formula:

$$\text{Number of color combinations} = \frac{7!}{(7-3)! \cdot 3!} = 35. \text{ The answer is B.}$$

## Solution to Problem 31

Because the securities need to be ranked, the order matters and we are dealing with permutations. Our task is to find the number of three-element ordered selections that can be created from the pool of 7 elements:

$$\text{Number of top-three lists} = \frac{7!}{(7-3)!} = 210$$

The answer is C.

## Solution to Problem 32

Because changing the order of letters creates a new arrangement, the order matters and we are dealing with permutations. We can find the number of rearrangements that can be created from a pool of 5 elements, 2 of which are the same (two T's), using the permutations formula for sets with repeating elements:

$$\text{Number of 5-letter combinations} = \frac{5!}{2!} = 3 \cdot 4 \cdot 5 = 60.$$ The answer is C.

## Solution to Problem 33

Since each number in the code must be a single-digit prime, the suitable digits are 2, 3, 5 and 7. Since no digit can be repeated, we have 4 elements to fill the 3 spots and can find the number of different codes from the permutations formula:

$$\text{Number of ID codes} = \frac{4!}{(4-3)!} = 24.$$ The answer is B.

## Solution to Problem 34

On problems that involve selections at multiple stages, it is often helpful to consider each step in the decision process individually and then compute the number of possibilities for each step and the total number of possible combinations.

1. Pizza flavor – 4 possibilities
2. Extra cheese – 2 possibilities (extra cheese/no extra cheese)
3. Mushrooms – 2 possibilities (mushrooms/no mushrooms)

Since extra cheese or mushrooms can be added to any flavor of the pizza, we can find the total number of varieties by multiplying the combinations for each of the 3 steps:

Total number of varieties = 4 · 2 · 2 = 16

The answer is D.

COMBINATORICS

## Solution to Problem 35

From the given percentages, we know that there are 3 novels, 2 study guides and 3 textbooks. Since each book is different, switching the order creates a new arrangement and we are dealing with permutations. Further, note that since we have to display all items, the new arrangements can be created only by rearranging books. The number of rearrangements in each group can be found as the factorial of the number of elements:

1.  Number of rearrangements for <u>novels</u>:        $3! = 6$
2.  Number of rearrangements for <u>study guides</u>:   $2! = 2$
3.  Number of rearrangements for <u>textbooks</u>:      $3! = 6$

Note that we cannot simply multiply all the arrangements in each group to get the final number of possibilities. This would be the case if we could not move the categories of books on the shelf, i.e. if, for example, the novels had to appear first followed by the study guides and textbooks. Since we are not precluded from moving the categories of books on the shelf, we can create new arrangements by rearranging the groups. Since we have 3 categories, let's find the number of ways to move them around:

Number of arrangements of the 3 categories of books $= 3! = 6$

Total number of book arrangements $= 6 \cdot 2 \cdot 6 \cdot 6 = 2 \cdot 6^3 = 2 \cdot 216 = 432$

The answer is E.


## Solution to Problem 36

First, let's find out how many different arrangements of students are possible. Since we can only reshuffle students among the 5 seats, there are a total of 5! or 120 arrangements.

Since Bob and Lisa must have only one neighbor from the group, they need to occupy seats 1 and 5 or seats 5 and 1, respectively. The remaining 3 students can take any of the 3 seats left in the middle, creating a total of 3! or 6 arrangements.

We can also sit Bob and Lisa in two ways (Bob in the left seat and Lisa in the right and vice versa) for each of the arrangements of the 3 students in the middle.

Number of arrangements with Bob and Lisa at the opposite ends $= 2 \cdot 3! = 12$

Total number of arrangements $= 5! = 120$

Probability to sit Bob and Lisa at the opposite ends $= \dfrac{12}{120} = 10\%$

The answer is B.

# PROBABILITY: BEATING THE ODDS

## HOW IT ALL STARTED

In an attempt to expand the scope of topics covered on the GMAT, the ETS first introduced probability in the problem solving section of the test in the early 1990's. After the adoption of the computerized test format in 1998, the test-makers have continued to increase the proportion and difficulty of the probability problems, infusing these concepts both into the problem solving and data sufficiency questions in the higher bins. As a result, the diversity of these questions dramatically increased, leading to a large number of non-trivial question groups combining probability with concepts from combinatorics, statistics, and number properties.

In this section, we will start with an overview of the major concepts and formulas you will need to know on the test, illustrating each with a pertinent test question. We will then proceed with the analysis of the strategies that are unique to probability questions and demonstrate their application to the upper-bin problems.

## CONCEPT OF PROBABILITY AND RANGE OF PROBABILITY VALUES

The concept of probability refers to the likelihood of the occurrence of a certain event. This likelihood can be measured in percents or in units of 1. If the event is certain to occur, its probability is said to be 1 (or 100%). For example, if you draw a ball from an urn that contains only white balls, the probability that this ball is white is 100% or simply 1. On the other hand, if the event cannot occur under the current circumstances, the probability of its occurrence is zero. Considering our example of a ball from an urn of white balls, the probability that this ball is black is zero because there were no black balls in the urn.

These two extreme cases denote boundaries for the values of probability – the minimum value of 0 and the maximum value of 1. In general, the entire spectrum of probabilities of any event falls between 0 and 1, inclusive. You can remember this intuitively; for instance, think about the fact that the probability of rain in a weather forecast will never be above 100% (or 1) or below 0% but will fluctuate in between these boundaries.

---

### TAKEAWAYS

---

1) Probability describes the likelihood of a certain event.

2) All probability values fall within the interval of [0; 1].

---

# PROBABILITY OF A SINGLE EVENT

The probability of a single event is determined by the ratio of the outcomes when this event occurs to the total number of possible outcomes of an experiment. For example, if an experiment involves making a blind guess on a GMAT question, we have a total of 5 outcomes of this experiment (choices A through E) but only one outcome gives us the correct answer. Thus, the likelihood of answering any GMAT question correctly based on a blind guess is one out of five ($\frac{1}{5}$) or 0.2 or 20%. In general, the probability of an event A is calculated as follows:

$$P(A) = \frac{\text{number of outcomes when A occurs}}{\text{total number of possible outcomes}}$$

To illustrate this concept, let's consider two relatively straightforward test problems that each involve finding the probability of a single event. At the higher levels of difficulty, the number of cases when the desired event occurs, as well as the total number of possible outcomes, will not be explicitly given to you. However, you will be able to compute these values based on the information in the problem, as illustrated in the two examples that follow. To give you a chance to solve these problems by yourself, solutions to the problems are provided separately on the following pages.

**Problem 37**                                    **Difficulty Level: 2**

Mathematics, physics, and chemistry books are stored on a library shelf that can accommodate 25 books. Currently, 20% of the shelf spots remain empty. There are twice as many mathematics books as physics books and the number of physics books is four greater than that of the chemistry books. If one book is randomly selected from the shelf, what is the probability that this is a chemistry book?

(A)  $\dfrac{1}{25}$

(B)  $\dfrac{2}{25}$

(C)  $\dfrac{1}{10}$

(D)  $\dfrac{3}{20}$

(E)  $\dfrac{9}{10}$

**Problem 38**                                    **Difficulty Level: 4**

A set consists of all integers between 600 and 1999, inclusive. If a number is selected at random from this set, what is the probability that it is divisible by both 5 and 13?

(A)   $\dfrac{65}{1399}$

(B)   $\dfrac{13}{280}$

(C)   $\dfrac{1}{50}$

(D)   $\dfrac{3}{200}$

(E)   $\dfrac{21}{1399}$

PROBABILITY

## SOLUTIONS: PROBLEMS 37 AND 38

### Solution to Problem 37

To find the probability that the selected book is about chemistry, we will need to find the number of chemistry books and divide it by the total number of books on the shelf. Since 20% of the 25 available book slots are empty, the total number of books on the shelf equals $25 - 25 \cdot 20\% = 20$.

Let m, p, and c denote the number of mathematics, physics, and chemistry books, respectively. Therefore, $m + p + c = 20$. Since the number of physics books is four greater than that of the chemistry books, $p = c + 4$. Also, given the fact that there are twice as many mathematics books as physics books:

$$m = 2p = 2(c + 4) = 2c + 8$$
$$m + p + c = 2c + 8 + c + 4 + c = 20$$
$$4c + 12 = 20$$
$$c = 2$$

After we found that there are 2 chemistry books, we can compute the probability of selecting a chemistry book as a ratio of the number of chemistry books to the total of 20 books on the shelf:

P(selected book is a chemistry book) $= \dfrac{\text{number of chemistry books}}{\text{total number of books}} = \dfrac{2}{20} = \dfrac{1}{10}$.

Therefore, the answer is C.

Note that when constructing the equation in this problem, we expressed all values in terms of c (the number of chemistry books). By applying this approach, we were able to find the number of chemistry books directly from solving the equation. By contrast, if we expressed values in terms of other variables, for example in terms of p ($m = 2p$ and $c = p - 4$), solving the equation would yield only the number of physics books and we would have to make another step to find the number of chemistry books.

---

### STRATEGY TIPS

**When constructing equations, find the variable that will provide the shortest way to answering the question, express it in terms of others and solve for the unknown.**

---

PROBABILITY

## Solution to Problem 38

Since both 5 and 13 are prime, their least common multiple, i.e. the smallest number divisible by 5 and 13, is $5 \cdot 13 = 65$. Therefore, all numbers divisible by both 5 and 13 have to be multiples of 65 and our task is to find the probability of picking a multiple of 65 from the set. Based on the probability formula for a single event,

$$P(A) = \frac{\text{number of outcomes when A occurs}}{\text{total number of possible outcomes}}, \text{ we need to find the number of multiples of}$$

65 in the set and divide it by the total number of integers between 600 and 1999.

The first number in the set that is evenly divided by 65 is 650 (note that because 65 goes into 600 a little more than 9 times, the first number in the set divisible by 65 will be a product of 65 and 10 – the next integer after 9). By applying the same logic, we can determine that the last number in the set divisible by 65 is 1950 (we can see that 1999 divided by 65 is slightly larger than 30, implying that the largest multiple of 65 in the set has to be a product of 65 and 30). Thus, we need to find the number of multiples of 65 between $65 \cdot 10$ and $65 \cdot 30$. This number will be equal to the number of integers between 10 and 30, since each of those integers when multiplied by 65 will yield a new multiple of 65.

The number of integers between 10 and 30, inclusive, equals $30 - 10 + 1 = 21$. Similarly, the total number of all integers in the set between 600 and 1999 is $1999 - 600 + 1 = 1400$.

Thus, the probability of drawing a number divisible by both 5 and 13 is $\frac{21}{1400} = \frac{3}{200}$ and the answer is D.

---

### STRATEGY TIPS

To find the number of multiples of any number k in the set, find the smallest number in the set divisible by k and the greatest number in the set divisible by k and represent them as a · k and b · k, where a and b are integers that result from dividing these numbers by k. Then the number of multiples of k in the set will be given by the number of integers between a and b computed as b – a + 1

---

PROBABILITY

# MUTUALLY EXCLUSIVE EVENTS

In this section we will review the concept of mutually exclusive events and analyze the types of problems on this topic. While you will not be asked about the theoretical terms and definitions on the test, you will need to be familiar with these concepts and formulas to be able to solve problems. We will limit the review of the theory to the essentials that you will need to solve each type of probability questions.

> **Two events are called <u>mutually exclusive</u> if they can never occur together, i.e. occurrence of one event completely eliminates the probability of occurrence of the other.**

In other words, we can say that if one event occurs, the probability that the other will occur is zero or, alternatively, the probability of these two events occurring together is zero. For example, if you take the GMAT, you can receive only one score on a given test session. Thus, for example, "scoring 750" and "scoring 730" on the same session are mutually exclusive events. In other words, we can say that the probability of both of these events occurring together (i.e. the probability of a test-taker getting two different scores on the same test session) is zero. Intuitively, once your 750 score shows up on the screen after the test, you know for sure that you did not get 730 or any other score on this test session, i.e. the probability of any other score is zero.

Consider an outcome of buying an instant lottery ticket with a chance of winning 1 in 10, a worthy investment of your hard-working money. Since only one outcome can occur on a single lottery ticket (i.e. it either loses or wins a certain amount), all possible outcomes on this ticket represent mutually exclusive events. For instance, "the ticket wins $10", "the ticket wins $100", and "the ticket does not win" are all mutually exclusive events because the occurrence of one event automatically excludes the occurrence of all others. In other words, you know that one ticket cannot simultaneously win $10 and win $100, i.e. the probability of these two events occurring together is zero. Similarly, if you know that one event has occurred (you scratched the surface and found that the ticket did not win), the probability that that same ticket wins you $100 or any other amount is zero.

> **If events A and B are mutually exclusive, P(A and B) = 0.**

## COMPLEMENTARY EVENTS

**Events are called <u>complementary</u> if one and only one of them must occur.** Since only one of the complementary events can occur, all complementary events are mutually exclusive. In other words, events complement one another if one and only one of these events <u>must</u> occur.

If we have only two events and one of the two events is certain to occur, we can conclude that the probability of occurrence of either one or the other is 1 or 100%. For example, if our humble goal is to get a perfect score on the quantitative section, the event "get a perfect score" and "get any other score" are mutually exclusive, since the occurrence of one event excludes the possibility of the other on the same test date. Also, these two events are complementary since it is certain that we will get either a perfect score or some other score.

As another example, "getting heads" and "getting tails" on a single toss of a coin are complementary events. Clearly, getting heads excludes the possibility of getting tails on the same toss and one of these two events must occur, since we are certain to get either tails or heads as a result of the toss, unless at this point in the book we have harnessed enough mental power to make the coin land on its edge.

---

**Events A and B are complementary if one and only one of them <u>must</u> occur.**

**For complementary events, P(A and B) = 0 and P(A or B) = 1.**

---

# PROBABILITY OF ONE EVENT **OR** ANOTHER

## 1. GENERAL CASE

The probability of occurrence of either one of the two events A and B is the sum of probabilities of event A and event B less the probability of both of these events occurring together.

$$P(A \text{ or } B) = P(A) + P(B) - P(A \text{ and } B)$$

Suppose we are looking to estimate our probability of getting offers of admission from two schools, an east-coast school and a west-coast school. Let event E (the left circle) denote an offer of admission from the east-coast school and event W (the right circle) denote an offer of admission from the west-coast school. The middle area on the diagram represents an intersection of the offers from our schools, the coveted outcome when we receive both offers and can choose which school to attend.

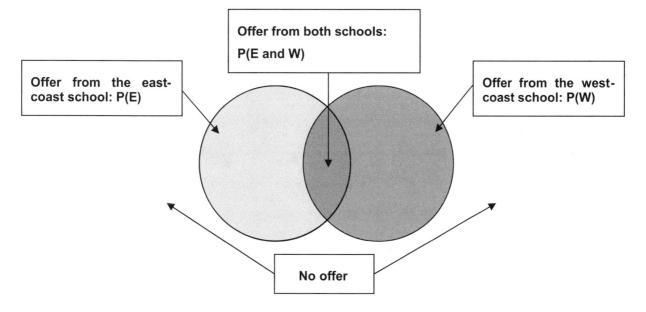

Since we can attend only one school, we would be satisfied to receive an offer of admission from either of the two reputable institutions and want to estimate the probability of this outcome. In other words, we would like to know P(E or W). The probability of an offer from either of the two schools is represented by the total area of the figure. Intuitively, the total area can be found as the sum of the <u>full</u> circle E and the <u>full</u> circle W less the area of the middle part where the two circles intersect.

Since the full left circle represents P(E) {an offer from the east-coast school}, the full right circle represents P(W) {an offer from the west-coast school), and the overlapping area denotes P(E and W) {offers from both schools}, we have :

$$P(E \text{ or } W) = P(E) + P(W) - P(E \text{ and } W)$$

PROBABILITY

97

Note that the probability of getting at least one offer (the area of the entire figure) includes the outcomes "offer from the east-coast school only" (left circle), "offer from the west-coast school only" (right circle) and "offer from both schools" (overlapping area). The counter-intuitive aspect is that the formula subtracts P(E and W), although "both E and W" represents a favorable outcome for our goal of at least one offer. The explanation for this counter-intuitive subtraction is that the probability P(E and W) denoted by the shared area is included in the full circle E as well as in the full circle W and gets counted twice when we add P(E) + P(W). Thus, P(E and W) is subtracted once to avoid double-counting.

Let's illustrate the application of this formula with an example and then apply it to the test problems.

## Exercise 12

Before taking the GMAT, you estimated that because of the strength of your academic record and professional background, you had a 40% chance to gain admission at the east-coast school and a 30% chance to land an offer from the west-coast school. You also estimated that the probability to land offers from both schools was 10%. After you scored in the 99th percentile on the test, you felt that your chances of getting offers from both schools improved: to 50% from the east-coast school and to 40% from the west-coast school. You also believe that the probability of getting offers from both schools increased from 10% to 15%. By what ratio did your GMAT performance increase your probability of landing an offer from at least one of the two schools?

## Solution

Your probability of landing an offer from at least one school is equivalent to the probability of landing an offer from the east-coast school, the west-coast school, or both. P(at least one offer) = P(E or W) = P(E) + P(W) – P(E and W). Let's now calculate the probabilities of at least one offer before and after the test.

Before the test: P(E or W) = 40% + 30% – 10% = 60%

After the test: P(E or W) = 50% + 40% – 15% = 75%

Ratio of probabilities = $\dfrac{75\%}{60\%} = 1.25$ .

Thus, your GMAT performance raised your chances of landing at least one offer from 60% to 75%, or by a ratio of 1.25.

Now, let's consider two representative GMAT problems of medium difficulty.

**Problem 39**                                        **Difficulty Level: 3**

Mathematics, physics, and chemistry books are stored on a library shelf that can accommodate 25 books. Currently, 20% of the shelf spots remain empty. There are twice as many mathematics books as physics books and the number of physics books is 4 greater than that of the chemistry books. Among all the books, 12 books are soft-cover and the remaining are hard-cover. If there are a total of 7 hard-cover books among the mathematics and physics books, what is the probability that a book selected at random is either a hard-cover book or a chemistry book?

(A)  $\dfrac{1}{10}$

(B)  $\dfrac{3}{20}$

(C)  $\dfrac{1}{5}$

(D)  $\dfrac{1}{4}$

(E)  $\dfrac{9}{20}$

**Problem 40**                                              **Difficulty Level: 3**

At a certain car dealership, the 40 vehicles equipped with air conditioning represent 80% of all cars available for sale. Among all the cars, there are 15 convertibles, 14 of which are equipped with an air-conditioning system. If a customer is willing to purchase either a convertible or a car equipped with air conditioning, what is the probability that a randomly selected vehicle will fit customer specifications?

(A) $\dfrac{3}{10}$

(B) $\dfrac{1}{3}$

(C) $\dfrac{9}{20}$

(D) $\dfrac{4}{5}$

(E) $\dfrac{41}{50}$

## SOLUTIONS: PROBLEMS 39 AND 40

### Solution to Problem 39

Let event C represent the outcome "the selected book is about chemistry" and event H denotes that "the selected book is hard-cover." In order to find P(C or H), from the formula P(C or H) = P(C) + P(H) – P(C and H), we know P(C), P(H), and P(C and H).

From problem 37, we know that the total number of books is 20 and the number of chemistry books is 2. Further, since among the 20 books all except 12 are hardcover, we know that the total number of hard-cover books is 8. Knowing that among the 8 hard-cover books, 7 are about mathematics and physics and the remaining are about chemistry, we can find the number of hard-cover chemistry books: 8 – 7 = 1. Now we can calculate all the necessary probabilities for our formula: selecting a chemistry book P(C), a hard cover book P(H), and a hard-cover chemistry book P(C and H).

$$P(C) = \frac{\text{number of chemistry books}}{\text{total number of books}} = \frac{2}{20} = \frac{1}{10}$$

$$P(H) = \frac{\text{number of hardcover books}}{\textit{total} \text{ number of books}} = \frac{8}{20} = \frac{2}{5}$$

$$P(C \text{ and } H) = \frac{\text{number of hardcover chemistry books}}{\textit{total} \text{ number of books}} = \frac{1}{20}$$

Therefore, P(C or H) = P(C) + P(H) – P(C and H) = $\frac{1}{10} + \frac{2}{5} - \frac{1}{20} = \frac{9}{20}$. The answer is E.

### Solution to Problem 40

Let C denote the event "the selected car is a convertible" and A – "the selected car is equipped with air-conditioning." To find P(C or A) from the formula P(C or A) = P(C) + P(A) – P(C and A), we must first compute P(C), P(A), and P(C and A).

Since 40 cars represent 80% of total vehicles, there are a total of 50 cars available for sale ($\frac{40}{0.8} = 50$). Because there are 15 convertibles,

$P(C) = \frac{\text{number of convertibles}}{\text{total number of cars}} = \frac{15}{50} = \frac{3}{10}$. Since 40 cars are equipped with an air-

conditioning system, $P(A) = \frac{\text{number of cars with an AC system}}{\text{total number of cars}} = \frac{40}{50} = \frac{4}{5}$. Given that 14

convertibles are AC-equipped, $P(C \text{ and } A) = \frac{14}{50} = \frac{7}{25}$.

Therefore, P(C or A) = P(C) + P(A) – P(C and A) = $\frac{3}{10} + \frac{4}{5} - \frac{7}{25} = \frac{41}{50}$. The answer is E.

PROBABILITY

## 2. MUTUALLY EXCLUSIVE EVENTS

The fact that mutually exclusive events cannot occur at the same time implies that P(A and B) = 0. If we use this property in the formula for the probability of at least one of the two events, P(A or B) = P(A) + P(B) – P(A and B), it follows that for mutually exclusive events A and B, the last term in the expression is zero and
P(A or B) = P(A) + P(B) – 0  = P(A) + P(B).

> **If A and B are mutually exclusive events, P(A or B) = P(A) + P(B).**

Let's illustrate this formula with an example.

## Exercise 13

A company's stock price depends on whether the firm meets its earnings estimates. The stock price will go up if the company exceeds its earnings expectations, remain the same if the company meets the earnings expectations, and decline if the firm fails to meet its earnings projections. The likelihood of each of the three scenarios for the next reporting period is 10%, 60%, and 30%, respectively. If the company issues a press release every time it **fails to meet** its earnings projections, but never when it **meets** its earnings projections, what is the probability that the company will **not** have to issue a press release for the next reporting period?

## Solution

The company will not have to issue a press release when it meets (event M) or beats (event B) its earnings estimates. The two events are mutually exclusive because the company's earnings cannot be equal to projections and be higher than these projections at the same time. Therefore, P(M or B) = P(M) + P(B) = 0.6 + 0.1 = 0.7 = 70%

Now, let's practice test problems on mutually exclusive events.

**Problem 41**                                          **Difficulty Level: 2**

In summer, Ajay sells ice-cream at a street stand. Each ice-cream bar, which costs Ajay 60 cents, is sold at the price of $1.50. Ajay can sell 200 ice-cream bars on a hot day, 100 bars on a cool day, and 50 bars on a cold day, scenarios occurring with the probability of 65%, 20% and 15%, respectively. What is the probability that Ajay's profits on a given day will exceed $67.58?

(A)   15%

(B)   65%

(C)   80%

(D)   85%

(E)   95%

**Problem 42**                                                **Difficulty Level: 4**

Set S consists of all non-negative even integers and all integers ending with 3 or 7. If an integer between 1 and 105, inclusive, is selected at random, approximately, what is the closest approximation to the probability that this integer is a member of set S?

(A)  50%

(B)  60%

(C)  70%

(D)  80%

(E)  90%

## SOLUTIONS: PROBLEMS 41 AND 42

### Solution to Problem 41

Since Ajay makes a profit of 90 cents on each ice-cream bar ($1.50 – $0.60), to make at least $67.58 in profit, he needs to sell at least 76 ice-cream bars (75 · $0.90 = $67.50 < 67.58 but 76 · $0.90 = $68.40 > $67.58). Thus, Ajay can meet his profit target only on a hot or cool day. A more elegant approach to arrive at this conclusion would be to estimate the profits that Ajay can make on a cold, cool, and hot day: $45, $90, and $180, respectively.

Since events "day is hot," "day is cool," and "day is cold" are mutually exclusive (e.g. one day cannot be hot and cold at the same time), then P("day is hot" or "day is cool") = P(day is hot) + P(day is cool) = 65% + 20% = 85% and the answer is D.

PROBABILITY

105

## Solution to Problem 42

Every 2nd number from 1 to 105 is even:    $\dfrac{105-1}{2} = 52$

$$P(\text{number is even}) = \dfrac{52}{105}$$

Every 10th number from 3 to 103 ends in a 3: $\dfrac{103-3}{10}+1 = 11$

$$P(\text{number ends with 3}) = \dfrac{11}{105}$$

Every 10th number from 7 to 97 ends in a 7: $\dfrac{97-7}{10}+1 = 10$

$$P(\text{number ends with 7}) = \dfrac{10}{105}$$

Since these events are mutually exclusive:
P(number is even or ends with 3 or 7) =
P(number is even) + P(number ends with 3) + P(number ends with 7) =
$\dfrac{52}{105} + \dfrac{11}{105} + \dfrac{10}{105} = \dfrac{73}{105} = 69.52\%$

Because 69.52% is closest to 70%, the answer is C.

Note that in order to find the probability of selecting an even number or a number ending in 3 or 7, we could simply compute the total number of integers fitting these descriptions and then divide this number by the total number of integers in the set, using the formula for a single event: $\dfrac{52+11+10}{105} = 69.5\%$. The reason we can simply add all favorable outcomes and divide them by the total number of outcomes is that mutually exclusive events can never occur together and simple counting of favorable outcomes of each event never results in double-counting. By contrast, if we were asked to find the total number of integers that are either even or divisible by five, we would not be able to simply add the number of even integers to the number of integers divisible by five because these events are not mutually exclusive. Since even integers ending with 0 are divisible by 5, adding the number of even integers to the number of multiples of five would result in double-counting of integers ending with 0.

---

### STRATEGY TIPS

**When finding the combined probability of several mutually exclusive events, simply add the number of outcomes of each event and divide by the total number of possible outcomes.**

---

### 3. COMPLEMENTARY EVENTS

The key property of complementary events tested on the GMAT is the fact that the sum of probabilities of all complementary events equals 1. Intuitively, you can remember this by recalling that one and only one of the complementary events <u>must</u> occur. Therefore, we have a 100% certainty that one of these events will happen. This 100% probability is then "partitioned" into probabilities of individual complementary events, implying that when we take the sum of these individual probabilities, we should get the same 100% again.

---

**If A and B are complementary events, P(A) + P(B) = 1.**

---

Let's illustrate this concept with a couple of problems.

## Problem 43                                    Difficulty Level: 1

When Rita goes bowling, she can knock down all the pins at one attempt with the probability of 10%, 9 pins with the probability of 10%, and 8 pins with the probability of 15%. Knowing her opponents, Rita estimates that she must knock down at least 8 pins in one attempt in order to win. What is the probability that Rita will NOT win?

(A)   10%

(B)   15%

(C)   35%

(D)   65%

(E)   75%

**Problem 44**                                    **Difficulty Level: 3**

An urn contains yellow, green, black, and orange balls. Among all the balls in the urn, 15% are yellow and 25% are green. If there are twice as many black balls as there are orange balls, what is the ratio of the probability of drawing a black ball to the probability of drawing either a yellow or a green ball?

(A)  $\dfrac{2}{10}$

(B)  $\dfrac{4}{15}$

(C)  $\dfrac{3}{5}$

(D)  1

(E)  2

# SOLUTIONS: PROBLEMS 43 AND 44

## Solution to Problem 43

Let's introduce the following notations of events:
  A – "Rita knocks down 10 pins";
  B – "Rita knocks down 9 pins";
  C – "Rita knocks down 8 pins".

We know that events A, B, and C are mutually exclusive because Rita can knock down only one number of pins at each attempt, i.e. if on a certain attempt the number of knocked pins is 8, it cannot be 9, 10, or any other number. In order to win, Rita needs to knock down 8, 9, or 10 pins. Then, P(Rita will win) = P(A or B or C) = 10% +10% + 15% = 35%  (the events are mutually exclusive). Events "Rita will win" and "Rita will not win" are also complementary events since one and only one of them must occur. Therefore, P(Rita will not win) = 1 – P(Rita will win) = 100% – 35% = 65%. The answer is D.

## Solution to Problem 44

If 15% and 25% of the balls are yellow and green, respectively, the remaining 60% of the balls (100% – 15% – 25%) are either black or orange, i.e. b + o = 60%. Since there are twice as many black balls as orange balls, i.e. $o = \dfrac{b}{2}$, then $b + \dfrac{b}{2} = 60\%$ and b = 40%.

Because 40% of balls are black, the probability of drawing a black ball is 40% (e.g. if there are a total of N balls, $0.40 \cdot N$ balls are black and the probability of selecting a black ball $P(B) = \dfrac{0.40 \cdot N}{N} = 40\%$). Similarly, the probability of drawing a yellow ball P(Y) is 15% and that of drawing a green ball P(G) is 25%.

To answer the question, we need to find $\dfrac{P(B)}{P(Y\,or\,G)}$. Because Y and G are mutually exclusive events (a ball cannot be yellow and green at the same time), P(Y or G) = P(Y) + P(G) = 15% + 25% = 40%. Then the ratio of P(B) to P(Y or G) = $\dfrac{P(B)}{P(Y\,or\,G)} = \dfrac{40\%}{40\%} = 1$. Therefore, the answer is D.

As you saw in the previous problem, it is not always necessary to know the exact number of outcomes and the exact total to compute the probability of a single event. We demonstrated that it is sufficient to know the proportion of the desired outcomes among the total number of outcomes. Thus, even if the exact numbers are unknown, we can compute probabilities using the ratio of desired outcomes to the total number of outcomes, a value that will appear on the GMAT as a percentage or a fraction.

### STRATEGY TIPS

**To compute the probability of a single event, it is sufficient to know the proportion of the outcomes when this event occurs among the total number of outcomes. It is not necessary to know the exact number of occurrences.**

PROBABILITY

# Probability of One Event **AND** Another

## 1. Dependent Events and Conditional Probability

So far we have considered only results of one experiment, such as tossing a coin once or drawing one ball from an urn. While many GMAT questions are limited to such one-round experiments, some of the higher-difficulty problems involve estimating a probability of a series of events. Examples of such questions include estimating the probability of getting balls of the same color in a number of sequential drawings from an urn or the likelihood of hitting the target in a series of attempts.

The probability of two events A and B occurring in sequence can be computed using the formula of conditional probability:

$$P(A \text{ and } B) = P(A) \cdot P_A(B)$$

Here P(A and B) is the probability that both these events will occur;
P(A) is the probability of event A;
$P_A(B)$ is the probability that event B will occur *assuming that A has already occurred*; $P_A(B)$ is called *conditional probability*.

To illustrate the concept of conditional probability, let's consider a few examples.

## Exercise 14

In a card game, a combination of two aces beats all others. If Jose is the first to draw from a standard deck of 52 cards, what is his probability of winning the game with the best possible combination?

## Solution

To win the game with the best possible combination, Jose needs to draw just two cards, each of which has to be an ace. To illustrate the notations in the formula of conditional probability, let event A denote "the first card is an ace" and event B denote "the second card is an ace." Since there are 4 aces among the 52 cards in the deck, the probability of drawing the first ace is $P(A) = \dfrac{4}{52} = \dfrac{1}{13}$. If the first card drawn by Jose is an ace, i.e. event A happens, then after Jose removes this card from the deck, there will be 51 cards remaining in the deck, among which 3 will be aces. Then the probability of drawing a second ace will be $P_A(B) = \dfrac{3}{51} = \dfrac{1}{17}$. In this case, $P_A(B)$ denotes the probability of event B ("the second card is an ace"), given that event A has already happened – the first card was an ace. Then the probability of drawing two aces in a row can be found from the formula of conditional probability: $P(A \text{ and } B) = P(A) \cdot P_A(B) = \dfrac{1}{13} \cdot \dfrac{1}{17} = \dfrac{1}{221} = 0.45\%$.

PROBABILITY

We can see that the probability that the second card is an ace depends on whether the first cards was or was not an ace. If the first card was not an ace, then after the first draw the deck will contain 51 cards but 4 rather than three of them will be aces. In this situation, the probability of drawing an ace at the second attempt will be $\dfrac{4}{51}$ (or about 7.8%) compared with $\dfrac{1}{17}$ (or about 5.9%) in the case when the first card was an ace. As you can see, the probability of the second event depends on the condition of whether the first event has occurred. Because of this interrelationship, events A and B are called **dependent events** and the probability of the occurrence of event B given that event A has occurred is given by the conditional probability $P_A(B)$.

> **Events A and B are <u>dependent</u> if the occurrence of one event affects the probability of another. The probability of dependent events occurring together is calculated using the general-case AND formula: $P(A \text{ and } B) = P(A) \cdot P_A(B)$**

Here $P(A \text{ and } B)$ is the probability that both these events will occur;
$P(A)$ is the probability of event A;
$P_A(B)$ is the probability that event B will occur *assuming that A has already occurred*;
$P_A(B)$ is called *conditional probability*.

Let's consider another example illustrating dependent events.

## Exercise 15

Derrick and Lena, a married couple attending the same business school, go to a corporate presentation that ends with two sequential drawings of a PDA (Personal Digital Assistant) among the 60 attending students. If each attendant is given one ticket participating in the lottery of the two PDAs and if each winning ticket is removed from the urn, what is the probability that both PDAs will go to the couple?

## Solution

In order for the couple to win the two PDAs, Derrick and Lena must each win a PDA. Let event A denote "the first PDA goes to the couple" and event B denote "the second PDA goes to the couple."

Since before the first drawing the couple has 2 lottery tickets from the 60 tickets participating in the drawing, the chance of one of the spouses to win the first PDA can be computed as usual – as a ratio of the number of favorable outcomes (one of the two tickets is selected and event A occurs) to the total number of outcomes in the drawing (60 tickets that can be selected). Thus, $P(A) = \dfrac{2}{60} = \dfrac{1}{30}$.

If one of the spouses wins the first PDA, i.e. event A occurs, the winning ticket is removed from the urn. Before the second drawing, there will be one PDA remaining to

allocate among the 59 remaining tickets. In other words, if event A occurs, the couple will have only one lottery ticket among the 59 tickets remaining in the run before the second drawing. Then the probability of event B given that event A has occurred will be $P_A(B) = \dfrac{1}{59}$. Therefore, the probability of both events A and B occurring together will be

$$P(A \text{ and } B) = P(A) \cdot P_A(B) = \frac{1}{30} \cdot \frac{1}{59} = \frac{1}{1770} = 0.056\%.$$

As we could see in the previous case, events A and B were dependent since winning the first PDA (and removing one of the couple's tickets from the urn) significantly reduced the chances of the couple to win the second PDA.

One of the most popular examples of dependent events on the GMAT includes drawing balls from an urn under the assumption that each selected ball is removed from the urn.

## Problem 45           Difficulty Level: 3

If 2 balls are randomly drawn from a green urn containing 5 black and 5 white balls and placed into a yellow urn initially containing 5 black and 3 white balls, what is the probability that the yellow urn will contain an equal number of black and white balls after this change?

(A)   $\dfrac{2}{9}$

(B)   $\dfrac{4}{9}$

(C)   $\dfrac{1}{2}$

(D)   $\dfrac{7}{10}$

(E)   $\dfrac{7}{8}$

Let's modify our example with the library books to illustrate a difficult test question on dependent events.

## Problem 46                                          Difficulty Level: 4

Mathematics, physics, and chemistry books are stored on a library shelf that can accommodate 25 books. Currently, 20% of the shelf spots remain empty. There are twice as many mathematics books as physics books and the number of physics books is 4 greater than that of the chemistry books. If 1 textbook selected at random is removed from the shelf to be shipped to another library, what is the probability of selecting a chemistry book among the remaining books?

(A)  $\dfrac{1}{190}$

(B)  $\dfrac{1}{19}$

(C)  $\dfrac{2}{19}$

(D)  $\dfrac{9}{95}$

(E)  $\dfrac{1}{10}$

## SOLUTIONS: PROBLEMS 45 AND 46

### Solution to Problem 45

Since no balls are taken out of the yellow urn and two balls are added to the urn, the total number of the balls in the yellow urn will increase from 8 to 10. Because the number of black balls must equal the number of white balls in the yellow urn, the number of white balls in this urn must increase from 3 to 5, implying that both balls added to the yellow urn from the green urn must be white. Thus, to answer the question, we need to determine the probability of selecting 2 white balls from the green urn under the condition that each of the selected balls is removed from the urn (balls are selected without replacement).

To illustrate the use of the conditional probability formula, let event A denote "the first ball taken out of the green urn is white," event B denote "the second ball taken out of the bowl is white." Then $P_A(B)$ is the probability of drawing a white ball at the second draw on the assumption that the first ball selected was also white (i.e. on the assumption that event A, "the first ball is white," has already occurred). Given that the green urn initially contained 5 white balls out of a total of 10, the probability of drawing a white ball at the first attempt is: $P(A) = \dfrac{5}{10} = \dfrac{1}{2}$.

Since we need to select two white balls in a row, we need to find the probability of selecting the second white ball given that 1 white ball has already been selected. After 1 white ball has been removed, the green urn will contain 4 white and 5 black balls among the 9 balls remaining in the urn. Thus, the probability to select a white ball at the second draw after 1 white ball has been removed is:

$P_A(B) = \dfrac{4}{9}$. Now that we know both $P(A)$ and $P_A(B)$, we can find $P(A \text{ and } B)$ – the probability that the first and the second ball removed from the green urn are white: $P(A \text{ and } B) = P(A) \cdot P_A(B) = \dfrac{1}{2} \cdot \dfrac{4}{9} = \dfrac{2}{9}$. The answer is A.

### Solution to Problem 46

From the prior library examples, we know that before the first book was shipped to another library, the total number of books on the shelf was 20, of which 2 were on chemistry and 18 were on other subjects – mathematics and physics. After one book is removed at random, there will be 19 books remaining on the shelf. However, the number of the available chemistry books will depend on whether the removed book was on chemistry or on another subject. If the removed book was on chemistry, there will be 1 chemistry book left. Otherwise, there will be 2 chemistry books remaining.

PROBABILITY

Let's define the outcomes of selecting the first book that is shipped to another library:

Event A1 – a chemistry book was removed;   $P(A1) = \dfrac{2}{20} = \dfrac{1}{10}$.

Event A2 – a book on physics or mathematics was removed;   $P(A2) = \dfrac{18}{20} = \dfrac{9}{10}$.

Note that events A1 and A2 are complementary since one and only one of them must occur – one of the books has to be shipped to another library and this book will be either on chemistry or on another subject but not both. Therefore, $P(A1) + P(A2) = 1$.

After 1 book was sent to another library, let event B denote selecting a chemistry book among the remaining 19 books. Since we do not know what book was shipped to another library, we need to find P(B) depending on whether event A1 or event A2 took place. In other words, we need to find the conditional probability of event B with respect to events A1 and A2. If event A1 occurred, there will be 1 chemistry book remaining and the probability of selecting a chemistry book in this case will be $P_{A1}(B) = \dfrac{1}{19}$.

Therefore, $P(A1 \text{ and } B) = P(A1) \cdot P_{A1}(B) = \dfrac{2}{20} \cdot \dfrac{1}{19} = \dfrac{2}{380}$.

If event A2 occurred, there will be 2 chemistry books left and $P_{A2}(B) = \dfrac{2}{19}$.

Therefore, $P(A2 \text{ and } B) = P(A2) \cdot P_{A2}(B) = \dfrac{18}{20} \cdot \dfrac{2}{19} = \dfrac{36}{380}$.

Finally, we can find the probability of selecting a chemistry book as the sum of probabilities of selecting a chemistry book under each of the two scenarios that represent mutually exclusive events (if scenario A1 happens, scenario A2 does not happen). Thus,

$P(B) = P(A1 \text{ and } B) + P(A2 \text{ and } B) = \dfrac{2}{380} + \dfrac{36}{380} = \dfrac{38}{380} = \dfrac{1}{10}$. The answer is E.

## 2. INDEPENDENT EVENTS

If you were not particularly excited about conditional probabilities in the library example, you will be glad to learn that the conditional probability matters only for dependent events – events for which the occurrence of one event affects the probability of another. A large portion of the GMAT problems on sequential events will involve only independent events, events whose occurrences do not depend on one another, a significantly simpler class of questions.

> **Events A and B are _independent_ if the occurrence of one event does not affect the probability of another.**

Examples of independent events commonly used on the GMAT include tossing a coin, throwing dice or drawing balls from an urn with replacement (returning each selected ball to the urn). The primary characteristic of independent events is that their outcomes are entirely independent of one another. In other words, to evaluate the probability of one event, we do not need to know anything about the occurrence of the other.

> **For independent events A and B, the probability of both events occurring is the product of the probabilities of the two events: (A and B) = P(A) · P(B)**

Note that this formula is a simplified version of the formula for dependent events. Since events A and B are independent and occurrence of one event does not affect the probability of another, the probability of B given the occurrence of A is the same as the simple probability of B: $P_A(B) = P(B)$.

Let's provide a few examples to illustrate common types of GMAT problems with independent events.

## Exercise 16

In a certain game of dice, the player's score is determined as a sum of three throws of a single die. The player with the highest score wins the round. If more than one player has the highest score, the winnings of the round are divided equally among these players. If Jim plays this game against 21 other players, what is the probability of the minimum score that will guarantee Jim some monetary payoff?

## Solution

Since we do not know scores of 21 other players, in order to <u>guarantee</u> that Jim will win some money this round, he must obtain the maximum possible score – a six on each of the three throws. Note that while it is possible for Jim to win with a lower score, only the maximum possible score will <u>guarantee</u> that Jim will be the winner of the round or, at least, that he will be among the winners of the round, if several people get the maximum score. Thus, to answer the question, we need to find the probability that Jim will get a six

on each of the three throws of dice. Since the dice have a form of cube, each integer between 1 and 6 is an equally likely outcome of a throw. Out of 6 possible outcomes of each throw (1, 2, 3, 4, 5, and 6), only one outcome will yield the desired result of 6.

Thus, the probability of getting a six on one throw = P(six on one throw) = $\dfrac{1}{6}$.

Note that the probability of each subsequent throw will not depend on the results of any other throws – it will be entirely independent. Intuitively, you can see this by realizing that regardless of whether you get 1, 2, 3 or any other number on the first throw, this will not increase or decrease your chances to get a 6 on the second throw because these throws are not linked to one another and the dice "will not remember" any of the prior results. You will always get a fresh start before each throw.

If we introduce events A, B, and C corresponding to getting a six on the first, second, and third throw, respectively, we know that events A, B, and C are independent and that P(A) = P(B) = P(C) = $\dfrac{1}{6}$. Since Jim needs a six on each of the three throws, all the events A, B, and C must happen. Thus, to find the probability of Jim getting the highest score, we need to compute P(A and B and C). Since the three events are independent, the probability that all three will occur can be found as a simple product of the probabilities of each of the three events: P(A and B and C) = P(A) · P(B) · P(C) = $(\dfrac{1}{6})^3$ = $\dfrac{1}{216}$.

Therefore, the probability that Jim will get a score that will guarantee him a win is $\dfrac{1}{216}$ or about 0.46%

---

### STRATEGY TIPS

---

**When a question asks about an outcome that <u>guarantees</u> a certain result, always consider the worst-case scenario to check that the condition will be satisfied even under the most unfavorable circumstances (e.g. Jim has to remain one of the winners even if all players get the highest possible score).**

---

Let's continue with the library example to demonstrate how GMAT problems are modified to test independent events.

## Problem 47                                       Difficulty Level: 3

Mathematics, physics, and chemistry books are stored on a library shelf that can accommodate 25 books. Currently, 20% of the shelf spots remain empty. There are twice as many mathematics books as physics books and the number of physics books is 4 greater than that of the chemistry books. Ricardo selects 1 book at random from the shelf, reads it in the library, and then returns it to the shelf. Then he again chooses 1 book at random from the shelf and checks it out in order to read at home. What is the probability Ricardo reads 1 book on mathematics and 1 on chemistry?

(A)   3%

(B)   6%

(C)   12%

(D)   20%

(E)   24%

While most test problems will ask you to compute the value of probability, some difficult questions will require you to estimate the actual quantities based on known probabilities. The next problem is a good example of this question type.

**Problem 48**                                    **Difficulty Level: 4**

Maria bought 4 black and a certain number of red and blue pencils at 15 cents each and carried them home in one bag. After Maria came home, she took out one pencil at random to write a note to her friend and then put this pencil back into the bag. Some time later, she needed to write another note and again took out a pencil from the bag. If the probability that Maria wrote both her notes in black is $\frac{1}{36}$, how much money did she spend on all pencils?

(A)    $0.90

(B)    $1.80

(C)    $3.60

(D)    $5.40

(E)    $7.20

## SOLUTIONS: PROBLEMS 47 AND 48

### Solution to Problem 47

Since Ricardo returned the first book to the shelf before selecting the second book, the two selections represent independent events. You can also see this from the fact that the first selection will have no effect on the second selection because regardless of the choice of the first book, this book is returned and mixed with the others, thus restoring the status quo.

Note that there are two possible scenarios for Ricardo to read 1 book on mathematics and 1 book on chemistry:
S1 – mathematics book in the library and chemistry book at home;
S2 – chemistry book in the library and mathematics book at home.

Since both of these scenarios will yield the desired outcome, we need to find the probability of S1 or S2. The probability of scenario S1 will be equal to the product of the probability that the first book is on mathematics by the probability that the second book is on chemistry (since the two events are independent), while the probability of scenario S2 will be equal to the probability that the first book is on chemistry and the second book is on mathematics. Knowing from earlier examples that there are 12 mathematics books and 2 chemistry books among the 20 available items before each selection, it follows that

$P(S1) = P(\text{first book is on mathematics}) \cdot P(\text{second book is on chemistry}) = \frac{12}{20} \cdot \frac{2}{20} =$

$\frac{24}{400} = \frac{6}{100} = 0.06 = 6\%$. Similarly, $P(S2) = P(\text{first book is on chemistry}) \cdot P(\text{second book}$

$\text{is on mathematics}) = \frac{2}{20} \cdot \frac{12}{20} = 0.06 = 6\%$

Finally, we know that scenarios S1 and S2 are mutually exclusive because, for example, if the first book is on mathematics in S1, S2 cannot occur because this scenario requires that the first book be on chemistry. Then using the OR formula for mutually exclusive events, $P(S1 \text{ or } S2) = P(S1) + P(S2) = 6\% + 6\% = 12\%$. The answer is C.

---

### STRATEGY TIPS

**If a question asks for the probability of an outcome that can occur under multiple scenarios, find the individual probabilities of each scenario. Then find the probability of the occurrence of one scenario OR any other scenario satisfying the conditions.**

## Solution to Problem 48

To find out the total amount spent on pencils, we need to find the total number of pencils purchased by Maria. Let N denote the total number of pencils. In order for both of Maria's notes to be written in black, both of the pencils that she used must have been black. Thus, we need to find the probability that both the first pencil selected and the second pencil selected were black – events A and B, respectively.

If Maria purchased a total of N pencils, four of which were black, then the probability of drawing the first black pencil at random is $P(A) = \dfrac{4}{N}$. Because she returned the first pencil to the bag after she finished her first note, the bag contained the same number of pencils of each color before she had to get another pencil for her second note. Then, the probability that the pencil used for the second note was black is $P(B) = \dfrac{4}{N}$.

Events A and B are independent because event A has no effect on event B. Intuitively, regardless of what pencil was selected for the first note, when it is returned to the bag and mixed with others, the initial situation is restored. Then the probability that Maria wrote both notes in black is $P(A \text{ and } B) = P(A) \cdot P(B) = \dfrac{4}{N} \cdot \dfrac{4}{N}$. Since we know from the problem that $P(A \text{ and } B) = \dfrac{1}{36}$, we can set up an equation $\dfrac{4}{N} \cdot \dfrac{4}{N} = \dfrac{1}{36}$. Solving for N, we get N = 24 or N = -24. Since the number of pencils must be positive, we know that Maria purchased 24 pencils. Given the 15-cent price of each pencil, the total amount paid was $15 \cdot 24 = \$3.60$. The answer is C.

---

### STRATEGY TIPS

**When asked for an actual number or quantity based on the probability of a certain event, introduce a variable to express the unknown and solve the equation.**

# STRATEGIES UNIQUE TO PROBABILITY

Now that we have covered the theoretical concepts and types of questions appearing on the probability section, we can concentrate on learning how to exploit the construction of the problems and answer choices to solve these problems in the most efficient manner.

## 1. THE LUCKY TWINS STRATEGY: IDENTIFYING PAIRS IN THE ANSWER CHOICES

This strategy allows us to exploit the construction of the answer choices on the GMAT to identify the likely answers even before solving the problem. This strategy exploits the design of the answer choices and is particularly effective with difficult questions on complementary events.

As we have seen in prior examples, many probability questions involve complementary events whose probabilities add up to one. Examples of such questions include finding the probability of winning the game (complementary to losing), the probability of getting at least one head in a coin toss (complementary to getting all tails), and the probability of hitting the target (complementary to missing the target).

Because the test authors construct answer choices with the intent of trapping students who almost get the question right, the answer choices for problems with complementary events very often include the probability of the event that is complementary to the one in the question stem. Knowing this property, you can identify the two answer choices that add up to one, one of which will be the correct answer and the other – the main distracter. By using other information from the problem, you will almost always be able to figure out which of the two is correct.

However, be careful with this strategy. Answer choices on hard questions may include two pairs of the Lucky Twins, making this approach difficult to implement. **You should use the Lucky Twins Strategy not to replace solving the problem but to make yourself aware of the traps when choosing the final answer.** Furthermore, this strategy is often helpful when you are forced to make a guess on a problem because of the time constraints.

Let's look at how this strategy can be applied to one of our earlier examples:

## Problem 43 revisited

When Rita goes bowling, she can knock down all the pins at one attempt with the probability of 10%, 9 pins with the probability of 10%, and 8 pins with the probability of 15%. Knowing her opponents, Rita estimates that she must knock down at least 8 pins in one attempt in order to win. What is the probability that Rita will NOT win?

(A)   10%
(B)   15%
(C)   35%
(D)   65%
(E)   75%

## Solution

Since the problem asks the probability that Rita will NOT win, the ETS will most certainly include an answer choice giving the probability that Rita WILL win to trap those test takers who missed the capitalized negative in the question. Since the event "Rita will not win" is complementary to the event "Rita will win," the probabilities of the two events must add up to one – the property you can use to spot this pair among other answer choices. By looking at the answer choices, you can see that choice C (35%) and choice D (65%) add up to 1. From the problem stem you can also see that because the probabilities of Rita knocking down 10, 9, and 8 bins are relatively small, she will probably be more likely to lose than to win. Knowing that the probability of losing must be higher than that of wining, you can select choice D (65%), the greater of the two complementary choices.

### When to use the Lucky Twins Strategy

The most common trigger phrase that should make you consider this strategy is "will NOT" in the problem stem. However, as will be shown in subsequent examples, this strategy can be applied to a broad range of other problems. In general, whenever you realize that the problem is intended to be solved though the use of complementary events, look at the answer choices to find a pair of probabilities that add up to 1. These answer choices will represent the probabilities of the two complementary events, from which you will be able to select the correct answer based on other information in the problem. If the occurrence of the event is more likely than non-occurrence, pick the greater of the two choices; otherwise, pick the smaller of the two. We will provide further illustrations of this strategy in the following problems.

## 2. THE ONE MIRROR STRATEGY: CONSTRUCTING COMPLEMENTARY EVENTS

Some GMAT problems will ask you to find the probability of an event that has a complicated structure of possible scenarios – for instance, getting 2 or 3 or 4 heads on 4 coin tosses. In this case, it is almost always more efficient to construct a complementary event, as finding its probability that will later enable you to find the probability of the desired event. We can find the probability of the desired event from its complementary event by subtracting the probability of the complementary event from 1.

The name of the strategy derives from the fact that probabilities of complementary events always add up to 1 and, like mirror images, complementary events can never exist without each other and never occur on the same side of the mirror (happen together).

### Problem 49                                          Difficulty Level: 2

If Jessica tosses a coin 3 times, what is the probability that she will get heads at least once?

(A)  $\dfrac{1}{8}$

(B)  $\dfrac{1}{4}$

(C)  $\dfrac{3}{8}$

(D)  $\dfrac{1}{2}$

(E)  $\dfrac{7}{8}$

## Solution to Problem 49

### Traditional approach

First, let's solve the problem using the traditional approach of evaluating possible scenarios. If we denote an outcome of heads on a single toss as H and tails as T, the possible scenarios that satisfy the condition in the problem are as follows:

1) getting heads once: HTT, THT, TTH,
2) getting heads twice: HHT, HTH, THH,
3) getting heads three times: HHH

The only scenario that does not satisfy the condition "heads appear at least once" is TTT. Therefore, out of a total of 8 possible outcomes of 3 tosses, 7 outcomes will result in heads appearing at least once. Then, the probability of getting heads at least once can be computed by dividing the number of outcomes when heads appear at least once by the total number of possible outcomes: $P = \dfrac{7}{8}$. The answer is E.

### The One Mirror Strategy in action

The previous approach required listing all possible outcomes, a methodology that is not only time consuming but also infeasible in cases with a larger number of possible scenarios (e.g. think about the outcomes of tossing a coin 10 times). A much more efficient way to solve problems involving multiple scenarios is to come up with an event that would be complementary to the one whose probability we need to calculate. Specifically, note that in this problem "getting heads at least once" is complementary to "getting tails three times" (the two can never happen together and one of the two events is certain to happen). Since the two events are complementary, P(getting heads at least once) = 1 – P(getting tails three times). Given that the probability of getting tails on one toss is $\dfrac{1}{2}$ and that each of the three tosses is an independent event,

$$P(\text{getting tails three times}) = \frac{1}{2} \cdot \frac{1}{2} \cdot \frac{1}{2} = \frac{1}{8}.$$

Then P(getting heads at least once) $= 1 - \dfrac{1}{8} = \dfrac{7}{8}$. The answer is E.

### The Lucky Twins Strategy in action

Note that we could have gotten the answer even more quickly using the Lucky Twins Strategy. By looking at the answer choices, we would immediately locate the only pair of choices that adds up to one – choice A ($\dfrac{1}{8}$) and choice B ($\dfrac{7}{8}$). Choice A will be given on a problem of this type to trap the test-takers who will do everything correctly but forget to do the last step in the calculation to get the probability of desired event from its complement ($1 - \dfrac{1}{8} = \dfrac{7}{8}$).

Also, you can deduce that the probability of getting heads on one toss is $\frac{1}{2}$ and the probability of getting heads <u>at least once</u> on several tosses is greater than $\frac{1}{2}$ (this probability increases with the number of tosses because we get more chances to end up with at least one head). Since you know that the probability of getting heads at least once is greater than $\frac{1}{2}$, the correct answer cannot be $\frac{1}{8}$ and must be $\frac{7}{8}$. The answer is E.

## Problem 50                                   Difficulty Level: 4

Set S consists of numbers 2, 3, 6, 48, and 164. Number K is computed by multiplying one random number from set S by one of the first 10 non-negative integers, also selected at random. If $Z = 6^K$, what is the probability that 678,463 is NOT a multiple of Z?

(A)   10%

(B)   25%

(C)   50%

(D)   90%

(E)   100%

## Solution to Problem 50

Since the vast majority of probability problems on the GMAT do not involve lengthy calculations, the stem of this question should prompt you to approach the problem analytically and look for shortcuts. Clearly, listing all possible values of K to find the probability is not feasible in this question. However, it is fairly easy to find the scenarios that will lead to the complementary event "678,463 <u>is</u> a multiple of Z." Because 678,463 is an odd number, it will be a multiple of $6^K$ only when K = 0, implying that $6^K = 6^0 = 1$ (all other non-negative integer values of K will result in an even number Z).

Since there are 5 numbers in set S, each of which can be multiplied by one of the ten non-negative integers, there are a total of $5 \cdot 10 = 50$ possible values of K. Out of the 50 possible values, only the products of each member of set S and 0 (i.e. $2 \cdot 0$, $3 \cdot 0$, $6 \cdot 0$, $48 \cdot 0$, and $164 \cdot 0$) will yield the desired zero value of K for which Z = 1. Since among the 50 possible values of K, only 5 will ensure that 678,463 is a multiple of Z, then

$$P(678{,}463 \text{ is a multiple of } Z) = \frac{5}{50} = 0.1.$$

Because we constructed an event complementary to the one in the problem stem, the probability that 678,463 is <u>not</u> a multiple of Z can be found by subtracting from 1 the probability that 678,463 <u>is</u> a multiple of Z: P(678,463 is <u>not</u> a multiple of Z) = 1 – P(678,463 is a multiple of Z) = 1 – 0.1 = 0.9 = 90%. The answer is D.

Note that we could have gotten the same answer by applying the Lucky Twins Strategy to the answer choices. Because choices A and D sum up to 1, we would anticipate that the right answer is either 10% or 90%. Since we know that 678,463 is an odd number, we would anticipate a high probability that this number will not be divisible by a power of an even number 6. Thus, between 10% and 90%, we would select 90%, getting the same answer as we the one we found by going through the calculations.

### When to use the One Mirror Strategy

This strategy is most effective on problems that that involve events with a large number of possible scenarios. These questions can be identified by trigger words "at least", "no more than" or "not happen" because such formulations typically identify a large scope of possibilities rather than just one possible outcome. For example, what is the probability that the sum of two dice will be *no more than* 10? (use the complementary event "the sum is more than 10" that has only two scenarios: "sum = 11" and "sum = 12"). What is the probability that at least one of the five balls selected from an urn with black and white balls is black? (Use the complementary event "all five selected balls are white.")

### 3. The Fresh Start Strategy: Differentiating Between Dependent and Independent Events

Since the differences between dependent and independent events are often subtle, this strategy will help you save time by quickly testing whether events are dependent or independent. In order to determine whether the events are independent, think about whether you are getting a fresh start before each event.

### Fresh Start Test

Let's consider a few of examples of applying the Fresh Start Test to the common types of events appearing on the test. If you throw dice five times and wish to maximize the outcome of each of throw, you will get a fresh start on each individual throw, regardless of how well or poorly you did on the others. You can also think that the die gets replaced after each throw and has no memory of what occurred in the prior attempts. Note that if you flip a coin a number of times, you will always get a fresh start on each toss. Similarly, if you draw balls from an urn and return each selected ball to the urn before drawing the next one, you will always get a fresh start before each subsequent drawing, regardless of whether you succeeded in selecting a ball of desired color in the prior attempt. Your record is cleared once you put the ball back into the urn.

Finally, if you throw five dice (or coins) simultaneously, you can think of these events as occurring in sequence to simplify the Fresh Start Test. For instance, if one die stops rolling before others and shows a five, you will still get a fresh start on other four dice that are still rolling (as they will have no memory of the outcome of the first die) – these events are therefore independent.

> In general, if the events pass the Fresh Start Test, they are independent and to find the probability of the events occurring together you can simply multiply their individual probabilities: $P(A \text{ and } B) = P(A) \cdot P(B)$

### Fresh Start Test Failed

By contrast, when the initial situation before any events is altered by the outcomes of the prior events, you cannot start fresh because you depend on what happened before. Under this scenario, the events are dependent.

For instance, if you draw a ball from an urn and remove it, the remaining number of balls and their composition will depend on the results of the prior drawings and you will not be able to start fresh before each new attempt. Similarly, if you select books from a library shelf and check them out, the person choosing the books after you will not be able to get a fresh start because some of the books will be missing. Finally, recalling the example about a corporate presentation with a drawing of two PDAs, if one of the spouses wins a PDA, the other spouse will not get a fresh start at the next drawing because the number of prizes will be reduced by one and the number of tickets in the drawing will also be reduced by one. Thus, all of the events that fail the Fresh Start Test are dependent and require the use of conditional probability formula: $P(A \text{ and } B) = P(A) \cdot P_A(B)$.

PROBABILITY

Let's illustrate how the Fresh Start approach can help you on the problems that involve both dependent and independent events:

## Problem 51                                    Difficulty Level: 4

Two identical urns – black and white – each contain 5 blue, 5 red and 10 green balls. Every ball selected from the black urn is immediately returned to the urn, while each ball selected from the white urn is removed and placed on a table. If Jenny receives a quarter for every blue ball, a dime for every red ball and a nickel for every green ball she selects, what is the probability that she will be able to buy a 25-cent candy bar with the proceeds from drawing four balls—two from each urn?

(A) $\dfrac{1}{16}$

(B) $\dfrac{9}{152}$

(C) $\dfrac{1}{2}$

(D) $\dfrac{25}{39}$

(E) $\dfrac{143}{152}$

## Solution to Problem 51

The fact that there are too many possible scenarios yielding the value of at least 25 cents should signal to you that you first need to use the One Mirror strategy to construct complementary events. The event complementary to "the proceeds are at least 25 cents" is the event "the proceeds are less than 25 cents." Since the only way for Jenny to make less than 25 cents on 4 balls is to draw 4 green balls, we need to find the probability of the scenario when both balls selected from the black urn and both balls selected from the white urn are green.

First, let's consider the black urn. Since the selected balls are returned to the urn after every drawing, Jenny will get a Fresh Start on each subsequent attempt. In other words, the fact that she selected a ball of a certain color in the prior drawing will have no effect on her future chances because each ball is returned and mixed with others, recreating the initial situation. Because Jenny gets a fresh start each time, we know that each of the drawings is independent. Since the initial situation is recreated before each drawing, we also know that the total number of balls and the number of green balls before each drawing is the same – 10 green balls among the 20 balls in total. Therefore, P(first ball from the black urn is green) = P(second ball from the black urn is green) = $\frac{10}{20} = \frac{1}{2}$.

Using the "AND" formula for independent events, P(both balls from the black urn are green) = P(first ball is green) $\cdot$ P(second ball is green) = $\frac{1}{2} \cdot \frac{1}{2} = \frac{1}{4}$.

Now, let's consider the white urn. Because each of the selected balls is removed from the urn, Jenny will never return to the initial number of balls; once she draws the first ball, she will not be able to get a fresh start on any subsequent drawing. Thus, the two drawings are dependent events and we should use the "AND" formula for dependent events: P(A and B) = P(A) $\cdot$ $P_A$(B). Since the urn initially contains 10 green balls, the probability that the first selected ball will be green is just $\frac{10}{20} = \frac{1}{2}$. After the first event has occurred (the first ball is green), the white urn will contain 9 green balls among the 19 balls remaining in the urn. Then P(the second ball is green assuming that the first ball is green) = $\frac{9}{19}$. By plugging these values into the "AND" formula for dependent events, we get that P(both balls selected from the white urn are green) = $\frac{1}{2} \cdot \frac{9}{19} = \frac{9}{38}$.

Finally, since we are using the two urns independently, the probability that both balls from the black urn are green AND that both balls from the white urn are green can be computed using the AND formula for independent events:

P(A and B) = P(A) $\cdot$ P(B) = $\frac{1}{4} \cdot \frac{9}{38} = \frac{9}{152}$.

Now we can compute the probability of generating the proceeds of at least 25 cents from the probability of the complementary event – the proceeds are less than 25 cents.

Specifically, P(proceeds are at least 25 cents) = 1 – P(proceeds are less than 25 cents) = $1 - \dfrac{9}{152} = \dfrac{143}{152}$, indicating that the correct answer is E.

Note that if we were running out of time on the test, we could quickly get the likely answer by using the Lucky Twins Strategy. We would identify the pair of answer choices $\dfrac{9}{152}$ and $\dfrac{143}{152}$ and then select the greater of the two to reflect the high likelihood of getting at least 25 cents from 5 proceeds.

# CHAPTER SUMMARY

## 1. CONCEPTS AND FORMULAS

**Probability** describes the likelihood of a certain event; all probability values fall between 0 and 1, inclusive: $0 \le P(A) \le 1$.

**Probability of a single event**: $P(A) = \dfrac{\text{number of outcomes when A occurs}}{\text{total number of possible outcomes}}$

---

**Mutually exclusive** events – events that can never occur together, i.e. occurrence of one event completely eliminates the probability of occurrence of the other.

Examples:

(1) Getting an A , a B, or a C as the final grade on one class.
(2) Answering 17 questions right or answering 21 questions right on a particular exam.

Note that the mutually exclusive events do not necessarily cover the entire scope of possible outcomes. For instance, it is also possible to get a D or an F as the final grade in the class in example 1, as well as it is possible to answer 19, 20, 25, etc. questions on a particular exam in example 2. Since the selected mutually exclusive events do not have to cover all possible outcomes, their probabilities do not have to add up to 1 – they are less than or equal to 1.

Important properties:

(1) Mutually exclusive events never occur together: $P(A \text{ and } B) = 0$
(2) $P(A \text{ or } B) = P(A) + P(B)$

---

Two events are **complementary** if one and only one of them must occur. All complementary events are mutually exclusive (because only one of them can occur) but not all mutually exclusive events are complementary.

Examples:

(1) Failing vs. passing a certain course (contrast with getting a particular grade)
(2) Answering at least half of the questions on the exam vs. answering less than half the questions on the exam (contrast with answering a particular number of questions in the example of mutually exclusive events).

Important properties:

(1) Complementary events never occur together: $P(A \text{ and } B) = 0$
(2) Because one of the complementary events must occur, their probabilities sum up to 1: $P(A \text{ or } B) = 1$; $P(A) = 1 - P(B)$; $P(B) = 1 - P(A)$.

PROBABILITY

Events A and B are **dependent** if the occurrence of one event affects the probability of another.

Examples:

(1) Drawing balls without replacement.
(2) Allocating a limited number of prizes among the audience (each subsequent allocation of a prize reduces the probability of the remaining participants winning).

$P(A \text{ and } B) = P(A) \cdot P_A(B)$

Where $P(A \text{ and } B)$ is the probability that both events will occur;
$P(A)$ is the probability of event A;
$P_A(B)$ is the probability that event B will occur *assuming that A has already occurred*;
$P_A(B)$ is called *conditional probability*.

---

Events A and B are **independent** if the occurrence of one event does not affect the probability of the occurrence of another. In case of sequential events, the initial situation is restored before each subsequent experiment.

Examples:

(1) Throwing dice
(2) Tossing a coin
(3) Selecting balls from an urn with replacement

Properties:

$P(A \text{ and } B) = P(A) \cdot P(B)$

---

**OR formulas** (Probability of A or B):
  (1) General case: $P(A \text{ or } B) = P(A) + P(B) - P(A \text{ and } B)$
  (2) Mutually exclusive events: $P(A \text{ or } B) = P(A) + P(B)$
  (3) Complementary events: $P(A \text{ or } B) = 1$

**AND formulas** (Probability of A and B):
  (1) Mutually exclusive and complementary events: $P(A \text{ and } B) = 0$
  (2) Dependent events: $P(A \text{ and } B) = P(A) \cdot P_A(B)$
  (3) Independent events: $P(A \text{ and } B) = P(A) \cdot P(B)$

PROBABILITY

## 2. STRATEGIES AND TIPS

- To compute the probability of a single event, it is sufficient to know the proportion of the outcomes when this event occurs among the total number of outcomes. It is not necessary to know the exact number of occurrences.

- When finding the combined probability of several mutually exclusive events, simply add the number of outcomes of each event and divide by the total number of possible outcomes.

- When asked about the probability of an outcome that <u>guarantees</u> a certain result, always consider the worst-case scenario as a check.

- When asked for an actual number or quantity based on the probability of a certain event, introduce a variable to express the unknown and solve the equation.

- To identify potential answers on difficult questions involving complementary events, use the Lucky Twins Strategy by looking for a pair of answer choices that adds up to 1.

- Use the One Mirror Strategy to find probabilities of events with a large number of scenarios by constructing a complementary event.

- Use the Fresh Start Strategy to differentiate between dependent and independent events.

# PROBABILITY PROBLEM SET

**10 QUESTIONS, 20 MINUTES**

### Problem 52                                    Difficulty level: 3

According to a recent student poll, $\frac{5}{7}$ of the 21 members of the finance club are interested in a career in investment banking. If two students are chosen at random, what is the probability that at least one of them is interested in investment banking?

(A) $\dfrac{1}{14}$

(B) $\dfrac{4}{49}$

(C) $\dfrac{2}{7}$

(D) $\dfrac{45}{49}$

(E) $\dfrac{13}{14}$

## Problem 53

**Difficulty level: 3**

If 4 fair dice are thrown simultaneously, what is the probability of getting at least one pair?

(A)  $\dfrac{1}{6}$

(B)  $\dfrac{5}{18}$

(C)  $\dfrac{1}{2}$

(D)  $\dfrac{2}{3}$

(E)  $\dfrac{13}{18}$

# Problem 54

**Difficulty level: 3**

Operation "#" is defined as adding a randomly selected two-digit multiple of 6 to a randomly selected two-digit prime number and reducing the result by half. If operation "#" is repeated 10 times, what is the probability that it will yield at least 2 integers?

(A)  0%

(B)  10%

(C)  20%

(D)  30%

(E)  40%

PROBABILITY

137

Number N is randomly selected from a set of consecutive integers between 50 and 69, inclusive. What is the probability that N will have the same number of factors as 89?

(A)   10%

(B)   13%

(C)   17%

(D)   20%

(E)   25%

## Problem 56

**Difficulty level: 3**

Each year three space shuttles are launched, two in June and one in October. If each shuttle launch is known to occur without a delay in 90% of the cases and if the current month is January, what is the probability that at least one of the launches in the next 16 months will be delayed?

(A) $\dfrac{1}{27}$

(B) $\dfrac{3}{27}$

(C) $\dfrac{271}{1000}$

(D) $\dfrac{729}{1000}$

(E) $\dfrac{26}{27}$

**Problem 57**                                        **Difficulty level: 4**

Crowan throws 3 dice and records the product of the numbers appearing at the top of each die as the result of the attempt. What is the probability that the result of any attempt is an odd integer divisible by 25?

(A)    $\dfrac{7}{216}$

(B)    $\dfrac{5}{91}$

(C)    $\dfrac{13}{88}$

(D)    $\dfrac{1}{5}$

(E)    $\dfrac{3}{8}$

PROBABILITY

**Problem 58**                                    **Difficulty level: 4**

A telephone number contains 10 digits, including a 3-digit area code. Bob remembers the area code and the next 5 digits of the number. He also remembers that the remaining digits are not 0, 1, 2, 5, or 7. If Bob tries to find the number by guessing the remaining digits at random, the probability that he will be able to find the correct number in at most 2 attempts is closest to which of the following?

(A)  $\dfrac{1}{625}$

(B)  $\dfrac{2}{625}$

(C)  $\dfrac{4}{625}$

(D)  $\dfrac{25}{625}$

(E)  $\dfrac{50}{625}$

**Problem 59**                                    **Difficulty level: 4**

If number N is randomly drawn from a set of all non-negative single-digit integers, what is the probability that $\dfrac{5N^3}{8}$ is an integer?

(A)  20%

(B)  30%

(C)  40%

(D)  50%

(E)  60%

## Problem 60

**Difficulty level: 4**

The acceptance rate at a certain business school is 15% for first-time applicants and 20% for all re-applicants. If David is applying for admission for the first time this year, what is the probability that he will have to apply no more than twice before he is accepted?

(A)  20%

(B)  30%

(C)  32%

(D)  35%

(E)  40%

**Problem 61**                                        **Difficulty level: 4**

If a randomly selected positive single-digit multiple of 3 is multiplied by a randomly selected prime number less than 20, what is the probability that this product will be a multiple of 45?

(A) $\dfrac{1}{32}$

(B) $\dfrac{1}{28}$

(C) $\dfrac{1}{24}$

(D) $\dfrac{2}{32}$

(E) $\dfrac{2}{28}$

## SOLUTIONS: PROBABILITY PROBLEM SET

### Answer Key:

52. E  55. D  58. E  61. C
53. E  56. C  59. D
54. A  57. A  60. C

### Solution to Problem 52

The quickest way to solve this problem is to use the One Mirror strategy. The complementary event to "at least one of the two students is interested in investment banking" (let's call this event A) is "none of the two students is interested in investment banking" (event B). Then, in order for event B to occur, each of the selected students must not be interested in banking. Since we know that there are 21 students, among whom 6 are not interested in banking ($\frac{2}{7}$ of 21), we can compute the probability for event B:

$$P(B) = \frac{6}{21} \cdot \frac{5}{20} = \frac{2}{7} \cdot \frac{1}{4} = \frac{1}{14}$$

Since we constructed event B to be complementary to event A, we can easily find the probability of event A:

$$P(A) = 1 - P(B) = 1 - \frac{1}{14} = \frac{13}{14}$$

The answer is E.

### Solution to Problem 53

Since there are many ways to get 1 or 2 pairs, a more efficient way to solve this problem is to construct a complementary event using the One Mirror strategy. The event complementary to and mutually exclusive with "there is at least one pair" is "there are no pairs." In order to get no pairs, each die must yield a different number; therefore, we can compute the number of outcomes with no pairs by multiplying the number of choices for each die. Since no outcome can be repeated, we will have 6 choices for the first die, 5 choices for the second die, 4 choices for the third die and 3 choices for the fourth die:

Outcomes with no pairs = $6 \cdot 5 \cdot 4 \cdot 3$

Total possible outcomes = $6 \cdot 6 \cdot 6 \cdot 6$

Probability of getting no pair = $\frac{6 \cdot 5 \cdot 4 \cdot 3}{6 \cdot 6 \cdot 6 \cdot 6} = \frac{5}{18}$

PROBABILITY

145

Probability of getting at least 1 pair $= 1 - \dfrac{5}{18} = \dfrac{13}{18}$

Also note that you could have spotted the pair of likely suspects ($\dfrac{5}{18}$ and $\dfrac{13}{18}$) by using the Lucky Twins Strategy. The answer is E.

## Solution to Problem 54

This problem can be solved without any computational work. We know that a multiple of 6 is always even and that any two-digit prime number is odd. Adding an even number to an odd will always yield an odd integer.

Since an odd integer is not divisible by 2, reducing the result by half will never yield an integer. Thus, regardless of how many times the operation is repeated, the probability to get an integer is 0. The answer is A.

## Solution to Problem 55

Since 89 is divisible only by 1 and 89, we are looking for a number with only two factors, i.e. a prime. To find the probability of selecting a prime, we need to count the number of primes between 50 and 69 and then divide it by the total number of integers between 50 and 69, inclusive. The total number of integers between 50 and 69 = 69 − 50 + 1 = 20. Note that after we found that there are 20 integers, we can eliminate answers B and C because these percentages of 20 will not yield a whole number.

Now, let's count the number of primes between 50 and 69. The quickest way to do this is to write out the 20 integers and then cross out all even numbers, multiples of 3, and multiples of 5. We are left with 53, 59, 61 and 67. Then, we only need to check their divisibility by 7. There is no need to check whether the four numbers are divisible by 11, 13 and other primes because, for instance, 11 or 13 could go into these integers only about 5 times, and we have already eliminated all numbers between 50 and 69 divisible by smaller primes. Thus, we have 4 primes among the 20 integers.

Probability to select a prime $= \dfrac{4}{20} = 20\%$. The answer is D.

## Solution to Problem 56

First, we need to find the number of launches that will occur in the next 16 months. Since it is currently January, 12 months from today will also be January and 16 months from today will be May (1$^{st}$ month + 4 = 5$^{th}$ month). Thus, three shuttle launches (2 in June + 1 in October) will be made in within this period.

To find the probability that at least one launch will be delayed (event A), we can apply the One Mirror Strategy and construct the complementary event B "none of the three launches are delayed." We can then compute the probabilities using the AND formula and then applying the properties of complementary events:

$$P(B) = P(\text{first launch on time}) \cdot P(\text{second launch on time}) \cdot P(\text{third launch on time}) = \frac{9}{10} \cdot$$

$$\frac{9}{10} \cdot \frac{9}{10} = \frac{729}{1000}$$

$$P(A) = 1 - P(B) = 1 - \frac{729}{1000} = \frac{271}{1000}$$

The answer is C.

## Solution to Problem 57

For the product of the three dice to be an odd number, all of the dice must produce an odd number (1, 3 or 5). Additionally, since this product must be divisible by 25, at least 2 of the dice must yield 5. Thus, two dice must yield a 5 and the remaining dice can yield 1, 3 or 5.

If we mark the dice as first, second, and third, 7 outcomes will suit us:
5-5-5   5-5-3   5-3-5   3-5-5   5-5-1   5-1-5   1-5-5

Total number of outcomes is $6 \cdot 6 \cdot 6 = 216$

According to the definition, the probability to get one of the suitable outcomes is $\frac{7}{216}$.

The answer is A.

## Solution to Problem 58

In order for Bob to dial the correct number in at most two attempts, he must find it either in his first attempt or in his second attempt. Thus, there are two possible scenarios:

(1) The first attempt is successful and the number is found;
(2) The first attempt is unsuccessful AND the second attempt is successful.

To find these probabilities, we need to compute the total number of possibilities for the remaining two digits. Since these digits cannot be 0, 1, 2, 5 or 7, we have 5 choices left for each of the two digits (3, 4, 6, 8, or 9).

Total number of possibilities = $5 \cdot 5 = 25$

Probability (scenario 1) = Probability (first attempt is successful) = $\dfrac{1}{25}$

Probability (scenario 2) = Probability (first attempt is unsuccessful) · Probability (second attempt is successful) = $\dfrac{24}{25} \cdot \dfrac{1}{24} = \dfrac{1}{25}$

Note that if the first attempt is unsuccessful in scenario 2, we will be able to eliminate one combination of digits and will have 24 rather than 25 possibilities before the second attempt.

Probability to guess the number in at most 2 attempts = Probability (Scenario 1 OR Scenario 2) = $\dfrac{1}{25} + \dfrac{1}{25} = \dfrac{2}{25}$. We can rewrite this to fit the answer choices: $\dfrac{50}{625}$.

Therefore, the answer is E.

## Solution to Problem 59

Since 5 is not divisible by 8, in order for $\dfrac{5N^3}{8}$ to be an integer, $N^3$ must be divisible by 8.

Note that for $N^3$ to be divisible by 8, it is sufficient that N be an even integer – any even integer is a multiple of 2 and consequently $N \cdot N \cdot N$ is a multiple of $2 \cdot 2 \cdot 2 = 8$. N = 0 is a special case, but it also results in an integer – 0.

Since we are dealing only with non-negative single-digit numbers, there are 5 even integers (0, 2, 4, 6, 8) among the 10 numbers. We have all the information to compute the probability:

Probability of selecting an even integer = $\dfrac{5}{10} = \dfrac{1}{2} = 50\%$

The answer is D.

## Solution to Problem 60

In order to David to submit no more than 2 applications, he must be accepted either this year or next year. Thus, there are two possible scenarios:

(1) Admission granted this year;
    OR
(2) Admission declined this year AND admission granted next year

Using the OR and AND formulas, we can compute the probability of admission within the next 2 years in the following steps:

P(scenario 1) = 15%

P(scenario 2) = P(decline as a first-time applicant) · P(admission as a re-applicant) = (1-15%) · 20% = 17%

P(admission within 2 years) = P(scenario 1 OR scenario 2) = 15% + 17% = 32%

The answer is C.

## Solution to Problem 61

To be a multiple of 45, the number has to be divisible by 9 and 5. The single-digit positive multiples of 3 are 3, 6 and 9. Since none of them is divisible by 5, the prime number in the product must be a multiple of 5. Since the only prime divisible by 5 is 5, the prime number must equal 5.

Since we will not get any factors of 3 from the prime number 5, divisibility by 9 has to be ensured by the 3 multiples of 3. Because only 9 is divisible by 9, we know that the only pair that will satisfy the condition in the problem is 9 and 5.

Finally, to find the probability, we need to compute the total number of possible cases. If we write out the prime numbers less than 20 (they are 2, 3, 5 ,7 ,11, 13, 17 and 19), we can see that there are 8 of them. We also know that there are 3 possibilities for the single-digit multiple of 3. Therefore, there are 24 total possibilities. Since only two pairs satisfy the condition, the probability to get a multiple of 45 is $\frac{1}{24}$. The answer is C.

# STRATEGIES FOR DATA SUFFICIENCY

## RECENT TRENDS IN DATA SUFFICIENCY

So far, the focus of our analysis of statistics, combinatorics, and probability has concentrated on the problem-solving questions. While these topics tend to appear more frequently in problem-solving, their use in data sufficiency has increased and is likely to grow in the future.

The goal of this section is to prepare you for the hardest data sufficiency questions. We will demonstrate how the theoretical concepts that we learned earlier are applied to data sufficiency questions and build some powerful strategies to attack these problems. Because the data sufficiency problems on combinatorics, statistics and probability cover the same scope of theory as those on problem solving, we will focus our attention only on the strategies that are unique to data sufficiency.

## STRATEGIES AND TIPS

### 1. START EASY

You may have noticed that making a conclusion about the sufficiency of the first statement leaves you with exactly the same number of possible answers as making a conclusion about the sufficiency of the second statement. For instance, if you know that the first statement is sufficient, you have two possible answers – A or D. Similarly, if you know that the second statement is sufficient, you are left with B or D. If you know that one statement is insufficient, you also have the same number of the remaining answer choices regardless of which statement you have found insufficient (the choices are B, C, or E or A, C, or E).

Since the test writers know that most people read the first statement before the second statement, the first statement is often created to be more time consuming. Because a conclusion about the first statement brings you just as close to the right answer as the conclusion about the second statement, always start with a statement that looks easier. By deciding the sufficiency of one of the statements, you will immediately eliminate two or three answer choices. Further, by considering the easier statement, you will gain comfort with the question and may discover some additional insight, which will help you make a quicker decision on the more difficult statement.

---

### STRATEGY TIPS

---

**Always start with an easier statement. By making a conclusion about the easier statement you will eliminate two or three answer choices and may discover additional insights to the question.**

---

Before you move on to the problems on the next page, review all the previous chapter summaries.

**Problem 62**                                                 **Difficulty Level: 2**

In how many ways can Pedro arrange his diplomas on a wall, if all the diplomas have to be hung at the height of 1.5 meters from the floor?

    (1) Pedro can display any 3 of his 5 diplomas;
    (2) The wall is 2.5 meters wide and 3 meters tall;

(A) Statement (1) alone is sufficient, but statement (2) alone is not sufficient.
(B) Statement (2) alone is sufficient, but statement (1) alone is not sufficient.
(C) BOTH statements TOGETHER are sufficient, but NEITHER statement ALONE is sufficient.
(D) Each statement ALONE is sufficient.
(E) Statements (1) and (2) TOGETHER are NOT sufficient.

## Solution to Problem 62

If all the diplomas must be hung at the same height, they must appear in a row. In other words, the question is asking for the number of arrangements that can be created by placing Pedro's diplomas in one row.

From a quick glance at the statements, we can see that the second statement simply gives the dimensions of the wall and has nothing to do with combinatorics. Therefore, it is a good way to start our analysis. Because we do not know the number of diplomas from statement (2), it is insufficient and we are left with A, C, or E.

From statement (1), we know that our task is to calculate the number of ordered 3-element arrangements from a pool of 5 elements. Note that the order matters because each of the diplomas is distinct. From the permutations formula:

The number of ways to arrange the diplomas $= \dfrac{5!}{(5-3)!} = 5 \cdot 4 \cdot 3 = 60$

Note that it is not necessary to get the final answer; we just need to realize that the specific number of elements in the pool and the number of distinct elements in the selection will yield one value for the total number of arrangements. The answer is A.

## 2. Be BAD

This strategy will help you guess intelligently on difficult problems. To understand its logic, let's think like test-makers for a moment. When the test writers create difficult data sufficiency problems, their goal is to make sure that these problems can accurately differentiate the people who know the correct answer from those who guess on the problem. Thus, their task is to predict how an average student would guess on a problem and then to make sure that these guesses are unlikely to yield the right answer.

From teaching experience, there are two most frequent ways for students to react to a difficult data sufficiency question. First, if a student finds information that is difficult to interpret and does not see its connection to the question, the initial impulse is often to pick C, presuming that at least if we put together the complicated expressions in both statements, we may be able to get the answer. A second common reaction among the students is to pick E. The reason is that if a student does not know the theoretical concept tested on the question, the test-taker often misses the subtle insight provided by the information in each statement. As a result, the test-taker misses the crucial clues provided in the statements and concludes that both statements are insufficient. Another scenario leading to an E-guess occurs when a student is completely confused by a hard question and chooses E as an "I have no idea" answer.

As a result, in an effort to reduce the proportion of correct guesses on <u>difficult</u> data sufficiency problems, the test-makers frequently hide the correct answer in the remaining three choices – A, B, or D. The takeaway is that if you have to guess on a difficult data sufficiency problem, pick A, B, or D (BAD is an easy way to memorize this). Therefore, if you see a difficult data sufficiency problem, try to make a decision on at least one of the two statements. For instance, if you conclude that the second statement is insufficient, then the possible answers are A, C, or E, of which only A is BAD. In this case, your bet would be to choose A.

---

### STRATEGY TIPS

**On difficult data sufficiency problems, the correct answer is most likely to be A, B, or D.**

---

DATA SUFFICIENCY

## Problem 63                                          Difficulty Level: 4

What is the maximum number of arrangements in which N students can be seated in a row of N seats at a movie theater, if all students from the same college are to sit next to each other?

   (1)   All students come from three colleges, X, Y, and Z that sent 12, 10, and 9 students, respectively;

   (2)   N is a prime number between 30 and 40;

(A) Statement (1) alone is sufficient, but statement (2) alone is not sufficient.
(B) Statement (2) alone is sufficient, but statement (1) alone is not sufficient.
(C) BOTH statements TOGETHER are sufficient, but NEITHER statement ALONE is sufficient.
(D) Each statement ALONE is sufficient.
(E) Statements (1) and (2) TOGETHER are NOT sufficient.

## Solution to Problem 63

From the first look at the problem, statement (2) clearly looks simpler than statement (1) and we should focus on it first. Note that even if you are not exactly sure how to solve this constrained combinatorics problem, your knowledge of primes should be sufficient to conclude that statement (2) yields two possibilities for the total number of students: 31 or 37. Because changing the number of students will change the number of arrangements that can be created, this statement is insufficient and the answer must be A, C, or E. Moreover, since this is a difficult combinatorics question, the answer is more likely to be A than C or E. If at this point you are running out of time, you should choose A and move on.

From statement (1), we know the total number of students is equal to 31 and that they must be seated in three groups of 12, 10, and 9 students. Because this provides all the information about the number of the students and their sitting pattern, it will yield to an exact number of maximum arrangements. Thus, statement (1) is sufficient and the answer is A.

If you are curious about the actual solution:

Because all students are attending, we can create new arrangements only by reshuffling them (remember that there are N! ways to rearrange N elements). Then, we have 12!, 10!, and 9! ways to rearrange students of each college within their sitting group. Further, we have 3 sitting groups and therefore have 3! ways to arrange the three colleges in the row of the movie theatre (e.g. X, Y, Z; or Y, X, Z, etc.). Thus, the total number of arrangements that can be created will be equal to the product of all the possibilities for rearranging: $3! \cdot 12! \cdot 10! \cdot 9!$

Note, however, that in this case your Start Easy strategy, coupled with intelligent guessing on BAD problems, would yield the correct answer A.

### 3. No News Is Good News

If the two statements provide the same information in different form, or if after algebraic transformations one of the statements equals the other, then the correct answer is D or E. In other words, if you discover that both statements supply essentially the same information, there are only two possible scenarios: (1) each statement alone is sufficient; or (2) both statements taken together are insufficient. Note that if two statements supply equivalent information, then one of the statements cannot be sufficient if the other is insufficient, thus ruling out choices A and B. Furthermore, putting together the two identical statements will not provide you with any additional insight, eliminating choice C.

---

### STRATEGY TIPS

**If the two statements provide the same information in different form, or can be equivalently transformed into one another, the correct answer is D or E.**

---

### Problem 64 
### Difficulty Level: 3

Set A consists of all positive integers less than 100; Set B consists of 10 integers, the first four of which are 2, 3, 5, and 7. What is the difference between the median of Set A and the range of Set B?

   (1) All numbers in Set B are prime numbers;
   (2) Each element in Set B is divisible by exactly two factors;

(A) Statement (1) alone is sufficient, but statement (2) alone is not sufficient.
(B) Statement (2) alone is sufficient, but statement (1) alone is not sufficient.
(C) BOTH statements TOGETHER are sufficient, but NEITHER statement ALONE is sufficient.
(D) Each statement ALONE is sufficient.
(E) Statements (1) and (2) TOGETHER are NOT sufficient.

## Solution to Problem 64

Because a prime number is a number that has exactly two factors (1 and itself), the two statements supply the same information – that all elements in set B are primes. Thus, the answer must be D or E.

To answer the question, we need to know the median of set A, as well as the smallest and the largest values in set B to compute the range. Since we know each element in set A, we can find the median of the set. Further, because statements (1) and (2) tell us that all members of set B are primes, we know that 2 is the smallest element in set B.

However, knowing that set B consists of 10 elements is insufficient to find the largest number in set B, since we do not know anything about other primes included in set B. To avoid the trap, we cannot assume that set B consists of consecutive primes just because its first four elements are consecutive primes. The next three members in set B could be 11, 13, 17 or any other combination of primes, such as 31, 59, 83, etc. Because we do not know the greatest value in set B, we do not have enough information to find the range of set B. The answer is E.

## 4. MEMORY BLACKOUT

Another common trick used by the test-makers is to construct the second statement so that it nicely complements the first. Because the test makers know that most people start with the first statement and then proceed with the second, they may try to trip up the students who confuse it with the initial information in the question.

After we read the first statement and start with the second, the clues from the first statement are often so fresh in our minds that the initial impulse is to automatically incorporate this information into the second statement. You need to resist this impulse to avoid a premature conclusion about the sufficiency of the second statement.

Just like a hero of a soap opera, you must harness the ability of getting quick and repetitive cases of amnesia and consider each of the two statements strictly individually. When you analyze each statement by itself, you need to forget everything you learned from the other statement before you have to consider the two of them together.

### STRATEGY TIPS

**When considering each statement individually, forget the information you learned from the other statement.**

**Problem 65**                                    **Difficulty Level: 4**

If sets X and Y have an equal number of elements, does set X have a greater standard deviation than Set Y?

   (1) The difference between each pair of the neighboring elements is consistent throughout each set;
   (2) Each of the first two elements in Set Y is twice greater than the corresponding first and second elements in Set X.

(A) Statement (1) alone is sufficient, but statement (2) alone is not sufficient.
(B) Statement (2) alone is sufficient, but statement (1) alone is not sufficient.
(C) BOTH statements TOGETHER are sufficient, but NEITHER statement ALONE is sufficient.
(D) Each statement ALONE is sufficient.
(E) Statements (1) and (2) TOGETHER are NOT sufficient.

DATA SUFFICIENCY

## Solution to Problem 65

From statement (1), we know that the two sets are each evenly spaced. However, because we do not know how spread out they are, this statement alone is insufficient to compare standard deviations.

From statement (2), we know that the first element of the set Y is twice greater than the first element of set X and that the second element of set Y is twice greater than the second element in set Y. However, we do not know anything about other elements in these sets, i.e. we do not know whether this trend will hold with the rest of the terms. In this case, it is crucial to forget that statement (1) tells us that each of the sets is evenly spaced. Otherwise, we may erroneously conclude that statement (2) is sufficient.

If we consider the two statements together, we know that set Y is a scaled-up version of set X. Because the difference between each pair of elements within the set is locked (from statement (1)), the pattern established by the first two elements (statement (2)) will hold for the remainder of the set. Thus, we can conclude that each of the elements in set Y is twice greater than the corresponding element in set X. Set Y will be more spread out, and because it has the same number of elements as set X, it will have a greater standard deviation.

For an intuitive explanation, recall also our Map Strategy from the Statistics section. Increasing (in this case doubling) the scale of the map will increase the standard deviation of the set. Therefore, two statements taken together are sufficient to answer the question and the answer is C.

## 5. Probability Ratios – As Good As It Gets

On probability questions, knowing the ratio of the cases when the event happens to those when the event does not happen is sufficient to compute the likelihood of the event. Note that it is not necessary to know the actual values or even the breakdown of cases when the event does not happen.

For example, if you would like to know the probability of selecting a green ball from an urn, it is sufficient to know the ratio of green balls to the total number of balls or the percentage of the green balls in the urn. It is not necessary to know the specific number of green balls or the total number of balls. More importantly, it is not even necessary to know what other kinds of balls are present in the urn (whether there are balls of 5 or 12 different colors other than green).

### Problem 66                                    Difficulty Level: 3

If a pencil is selected at random from a desk drawer, what is the probability that this pencil is red?

  (1) There are 6 black and 4 orange pencils among the pencils in the drawer;
  (2) There are three times as many red pencils in the drawer as pencils of all other colors combined;

(A) Statement (1) alone is sufficient, but statement (2) alone is not sufficient.
(B) Statement (2) alone is sufficient, but statement (1) alone is not sufficient.
(C) BOTH statements TOGETHER are sufficient, but NEITHER statement ALONE is sufficient.
(D) Each statement ALONE is sufficient.
(E) Statements (1) and (2) TOGETHER are NOT sufficient.

### Solution to Problem 66

Because statement 1 provides no information about the number of red pencils, it is insufficient. The answer must be B, C or E.

From statement 2, we know that red pencils are three times as prevalent as those of any other color. Thus, if there are x non-red pencils, then there are 3x red pencils and 4x pencils in total. From this information alone, we can compute the probability of selecting a red pencil.

Thus, statement (2) alone is sufficient and the answer is B.

$$\text{Probability of selecting a red pencil} = \frac{\text{number of red pencils}}{\text{total number of pencils}} = \frac{3x}{4x} = \frac{3}{4} = 75\%$$

---

### STRATEGY TIPS

**On probability questions, the ratio of the cases when the event happens to those when the event does not happen is sufficient to compute the probability. It is not necessary to know the actual values or even the breakdown of cases when the event does not happen.**

---

## STRATEGY SUMMARY

- Always start with an easier statement. By making a conclusion about an easier statement, you will eliminate two or three answer choices and may discover additional insights to the question.

- On difficult data sufficiency problems, the correct answer is more likely to be A, B, or D.

- If the two statements provide the same information in different form or can be equivalently transformed into each other, the correct answer is D or E.

- When considering each statement individually, forget the information you learned from the other statement.

- On probability questions, the ratio of the cases when the event happens to those when the event does not happen is sufficient to compute the probability. It is not necessary to know the actual values or even the breakdown of cases when the event does not happen.

# DATA SUFFICIENCY PROBLEM SET

**10 QUESTIONS, 20 MINUTES**

**Problem 67**                                          **Difficulty level: 2**

What is the probability of selecting a white ball from an urn?

   (1) There are twice as many white balls as there are balls of any other color
   (2) There are 30 more white balls as balls of all other colors combined

(A) Statement (1) alone is sufficient, but statement (2) alone is not sufficient.
(B) Statement (2) alone is sufficient, but statement (1) alone is not sufficient.
(C) BOTH statements TOGETHER are sufficient, but NEITHER statement ALONE is sufficient.
(D) Each statement ALONE is sufficient.
(E) Statements (1) and (2) TOGETHER are NOT sufficient.

**Problem 68**                                   **Difficulty level: 2**

At a business school conference with 100 attendees, are there any students of the same age (rounded to the nearest year) who attend the same school?

(1) The range of ages of the participants is 22 to 30, inclusive
(2) Participants represent 10 business schools

(A) Statement (1) alone is sufficient, but statement (2) alone is not sufficient.
(B) Statement (2) alone is sufficient, but statement (1) alone is not sufficient.
(C) BOTH statements TOGETHER are sufficient, but NEITHER statement ALONE is sufficient.
(D) Each statement ALONE is sufficient.
(E) Statements (1) and (2) TOGETHER are NOT sufficient.

**Problem 69**                                               **Difficulty level: 2**

Jonathan would like to visit one of the 12 gyms in his area. If he selects a gym at random, what it the probability that the gym will have both a swimming pool and a squash court?

  (1) All but 2 gyms in the area have a squash court
  (2) Each of the 9 gyms with a pool has a squash court

(A) Statement (1) alone is sufficient, but statement (2) alone is not sufficient.
(B) Statement (2) alone is sufficient, but statement (1) alone is not sufficient.
(C) BOTH statements TOGETHER are sufficient, but NEITHER statement ALONE is sufficient.
(D) Each statement ALONE is sufficient.
(E) Statements (1) and (2) TOGETHER are NOT sufficient.

**Problem 70**                                    **Difficulty level: 2**

How many ways does a coach have to select a university team from a pool of eligible candidates?

   (1) The number of eligible candidates is 3 times greater than the number of slots on the team
   (2) 60% of the 20 athletes are eligible to play on the 4-person university team

(A) Statement (1) alone is sufficient, but statement (2) alone is not sufficient.
(B) Statement (2) alone is sufficient, but statement (1) alone is not sufficient.
(C) BOTH statements TOGETHER are sufficient, but NEITHER statement ALONE is sufficient.
(D) Each statement ALONE is sufficient.
(E) Statements (1) and (2) TOGETHER are NOT sufficient.

**Problem 71**                                    **Difficulty level: 3**

What is the median of set A {-8, 15, -9, 4, N)?

  (1) Number N is a prime and $N^6$ is even
  (2) 2N + 14 < 20

(A) Statement (1) alone is sufficient, but statement (2) alone is not sufficient.
(B) Statement (2) alone is sufficient, but statement (1) alone is not sufficient.
(C) BOTH statements TOGETHER are sufficient, but NEITHER statement ALONE is sufficient.
(D) Each statement ALONE is sufficient.
(E) Statements (1) and (2) TOGETHER are NOT sufficient.

**Problem 72**                                          **Difficulty level: 3**

There were initially no black marbles in a jar. Subsequently, new marbles were added to the jar. If marbles are drawn at random and the selected marbles are not returned to the jar, what is the probability of selecting 2 black marbles in a row?

(1) After the new marbles are added, 50% of all marbles are black
(2) Among the 10 added marbles, 8 are black

(A) Statement (1) alone is sufficient, but statement (2) alone is not sufficient.
(B) Statement (2) alone is sufficient, but statement (1) alone is not sufficient.
(C) BOTH statements TOGETHER are sufficient, but NEITHER statement ALONE is sufficient.
(D) Each statement ALONE is sufficient.
(E) Statements (1) and (2) TOGETHER are NOT sufficient.

## Problem 73                       Difficulty level: 3

In how many ways can N students be seated in a row with N seats?

    (1) $|N - 6| = 3$
    (2) $N2 = 7N + 18$

(A) Statement (1) alone is sufficient, but statement (2) alone is not sufficient.
(B) Statement (2) alone is sufficient, but statement (1) alone is not sufficient.
(C) BOTH statements TOGETHER are sufficient, but NEITHER statement ALONE is sufficient.
(D) Each statement ALONE is sufficient.
(E) Statements (1) and (2) TOGETHER are NOT sufficient.

**Problem 74**                                              **Difficulty level: 4**

What is the probability that it will rain on each of the next 3 days if the probability of rain on any single day is the same in that period?

(1) The probability of no rain throughout the first two days is 36%
(2) The probability of rain on the third day is 40%

(A) Statement (1) alone is sufficient, but statement (2) alone is not sufficient.
(B) Statement (2) alone is sufficient, but statement (1) alone is not sufficient.
(C) BOTH statements TOGETHER are sufficient, but NEITHER statement ALONE is sufficient.
(D) Each statement ALONE is sufficient.
(E) Statements (1) and (2) TOGETHER are NOT sufficient.

## Problem 75

**Difficulty level: 4**

Set X consists of different positive numbers arranged in ascending order: K, L, M, 5, 7. If K, L, and M are consecutive integers, what is the arithmetic mean of set X?

    (1) The product K · L · M is a multiple of 6
    (2) There are at least 2 prime numbers among K, L and M

(A) Statement (1) alone is sufficient, but statement (2) alone is not sufficient.
(B) Statement (2) alone is sufficient, but statement (1) alone is not sufficient.
(C) BOTH statements TOGETHER are sufficient, but NEITHER statement ALONE is sufficient.
(D) Each statement ALONE is sufficient.
(E) Statements (1) and (2) TOGETHER are NOT sufficient.

**Problem 76**                                     **Difficulty level: 4**

If a number is drawn at random from the first 1,000 positive integers, what is the probability of selecting a *refined* number?

    (1) Any *refined* number must be divisible by 22
    (2) A refined number is any even multiple of 11

(A) Statement (1) alone is sufficient, but statement (2) alone is not sufficient.
(B) Statement (2) alone is sufficient, but statement (1) alone is not sufficient.
(C) BOTH statements TOGETHER are sufficient, but NEITHER statement ALONE is sufficient.
(D) Each statement ALONE is sufficient.
(E) Statements (1) and (2) TOGETHER are NOT sufficient.

## SOLUTIONS: DATA SUFFICIENCY PROBLEM SET

### Answer Key:

67. A  70. B  73. B  76. B
68. C  71. A  74. D
69. B  72. C  75. E

### Solution to Problem 67

From statement (1) we know that selecting a white ball is twice as likely as selecting a ball of any other color, which is sufficient to find the probability of selecting a white ball. If you would like to verify this, here is a quick illustration. If we denote the number of non-white balls as X, the number of white balls is 2X and the total number of balls is 3X.

$$\text{Probability to select a white ball} = \frac{2X}{3X} = \frac{2}{3}$$

Statement (1) alone is sufficient.

From statement (2) we know only the difference between the number of white balls and balls of other colors. This is insufficient to determine the actual number of balls or the ratio of white balls to total balls that would be necessary to find the probability. Intuitively, you can see that there can be 31 white balls and 1 ball of another color (probability = $\frac{31}{32}$) or 60 white balls and 30 balls of other colors (probability = $\frac{2}{3}$). Therefore, the second statement is insufficient and the answer is A.

### Solution to Problem 68

Statement (1) tells us that there are 9 age categories at the conference but does not provide us with any information about the participating schools. Thus, this statement is insufficient.

Statement (2) tells us that there are 10 categories for schools but gives no information about the age of the participants.

By combining the two statements together, we know that we can create a maximum of 90 distinct combinations of age/school (9 choices for age · 10 choices for school). Since we have 100 participants and only 90 different combinations, some of the attendees must have the same combination, i.e. must represent the same school and be of the same age. Therefore, the answer is C.

## Solution to Problem 69

To answer the question, we need to find the number of gyms that have both a pool and a squash court. Statement (1) does not tell us anything about the number of gyms with a pool and is therefore insufficient to answer the question.

From statement (2), we learn that there are 9 gyms with a pool and that each of them is equipped with a squash court. Therefore, there are 9 gyms that have both amenities. Note that the total number of gyms that have a squash court is not needed, since only those of them that have a pool will qualify. The answer is B.

## Solution to Problem 70

To find the number of team combinations, we need to know the number of eligible candidates and the number of slots on the team. Statement (1) simply gives us the relationship between these two values and is therefore insufficient to find the number of combinations. Intuitively, you can see that if there are 2 people on the team and 6 candidates, the number of combinations is much smaller than if there are 10 people on the team and 30 eligible candidates.

From statement (2), we know that we have 12 candidates to fill the 4 spots on the team; In other words, statement (2) provides us both with the size of the pool and the size of the selection, thus locking in one value for the number of possible combinations ($\frac{12!}{(12-4)! \cdot 4!}$). Therefore, statement (2) alone is sufficient and the answer is B.

---

### STRATEGY TIPS

**While ratios are often sufficient to compute probabilities, they are usually insufficient to find the number of permutations or combinations that can be created.**

---

## Solution to Problem 71

Since Set A contains an odd number of terms, the median of the set will be the number in the middle of the set, if the terms are arranged in ascending or descending order, i.e. the third largest number in the set. Thus, to answer the question, we need to find the third largest number in the set.

From statement (1), if $N^6$ is even, N must also be even. Since N is a prime number, we know that N = 2, the only even prime. Thus, the elements of Set A arranged in ascending order are {-9, -8, 2, 4, 15} and the median of the set is 2. Statement (1) alone is sufficient.

From statement (2), we know that N < 3. This information alone is insufficient to answer the question. For example, if N = 2, then the median of the set is 2, while if N = -10, then the elements of the set are {-10, -9, -8, 4, 15} and the median of the set is -8.
Therefore, the answer is A.

## Solution to Problem 72

To find the probability that the first selected marble is black, we need to know only the proportion of the black marbles in the jar after the new marbles are added. However, after the first black marble is selected, the number of all marbles (T) and the number of black marbles (B) will be each reduced by 1. In other words, the probability that the second marble will be black after the first marble was black will be equal to $\frac{B-1}{T-1}$. To find this probability, we need to know the exact values of the total number of marbles and the number of black marbles (for instance, $\frac{5-1}{10-1}$ is different from $\frac{50-1}{100-1}$).

Statement (1) does not provide us with any information about the exact number of marbles in the jar. While the ratio of the black marbles is sufficient to compute the probability that the first selected marble is black, it is insufficient to find the probability that the second selected marble is also black. Statement (1) alone is insufficient.

Statement (2) provides us with the information about the number of black marbles added but gives no information on the total number of marbles in the jar and is therefore insufficient.

If we combine the two statements, we know that the 8 black marbles in the jar constitute 50% of all marbles and that there are a total of 16 marbles in the jar. Since we know the total number of marbles and the number of black marbles, we can answer the question. Statements (1) and (2) taken together are sufficient to answer the question. The answer is C.

---

### STRATEGY TIPS

**While the ratios are usually sufficient to find simple probabilities, they are normally insufficient to compute conditional probabilities.**

---

## Solution to Problem 73

Since the number of seats is equal to the number of students, we need to find the number of ways in which the students can be rearranged within one row – the value equal to N! In other words, we simply need to find N.

From statement (1), we get two simple equations:
$|N - 6| = 3$
$N - 6 = 3$ or $N - 6 = -3$
$N = 9$ or $N = 3$

Since we have two solutions, both of which are positive and could represent the number of students, we cannot give a definitive answer to the question. Statement (1) alone is insufficient.

From statement (2), we get a quadratic equation:
$N^2 = 7N + 18$
$N^2 - 7N - 18 = 0$
$(N - 9)(N + 2) = 0$
$N = 9$ OR $N = -2$

Since the number of students cannot be negative, only $N = 9$ fits the problem stem. We have one value for the number of students and statement (2) is sufficient to answer the question. The answer is B.

## Solution to Problem 74

The probability that it will rain on each of the 3 days (day 1 AND day 2 AND day 3) is equal to the product of the individual probabilities of rain on each day (p1 · p2 · p3).
Since the probability of rain is the same for each of the 3 days and is not affected by the outcomes of other days (independent), the probability that it will rain every day is simply $p^3$, where p is the probability of rain on any single day. Thus, to answer the question we simply need to find the probability of rain on any single day.

Since statement (2) is simpler, this is a good start for your analysis. Statement (2) provides us with the information on the probability of rain on one single day and is therefore sufficient to answer the question.

Statement (1) provides us with the information about the probability of two dry days in a row. Since the events "a dry day" and "a rainy day" are mutually exclusive, the probability to have no rain on any single day is 1-p, where p is the probability of rain on any single day, as we noted earlier. Since we are given the information about the first and the second days being dry, we can construct an equation and solve it for p, thus answering the question:
P(no rain on day 1 AND no rain on day 2) = (1 - p)(1 - p) = 0.36
$$(1 - p)^2 = 0.36$$
$$p = 0.4$$

Each statement alone is sufficient to answer the question and the answer is D.

## Solution to Problem 75

To find the mean of the set, we need to determine the values of K, L and M or their sum. Since K, L and M have to be consecutive positive integers less than 5 (the set is arranged in ascending order and includes only different terms), K, L and M can be either 1, 2 and 3 or 2, 3 and 4.

From statement (1), we know that both 2 · 3 · 4 and 1 · 2 · 3 will be multiples of 6. Note that any product of 3 positive consecutive integers is a multiple of 6. Since there is at least one even number among any 3 consecutive integers, their product will always be divisible by 2. Further, since 3 consecutive integers go at the step of 3, one of them will

always be a multiple of 3 and their product will always be divisible by 3. Since the product is divisible by 2 and 3, it will be divisible by 6. Statement (1) provides no new information about K, L and M and is therefore insufficient.

From statement (2), we know that K, L and M can either be 2, 3 and 4, or 1, 2, and 3, since both sequences contain two prime numbers. Statement (2) provides no new information about K, L and M and is therefore insufficient. The answer is E.

---

### STRATEGY TIPS

**The product of any 3 consecutive positive integers is always a multiple of 2, 3 and 6.**

---

## Solution to Problem 76

Statement (1) tells us that a refined number must be a multiple of 22. Note, however, that this is a necessary but not sufficient condition for a refined number. In other words, while every refined number is a multiple of 22, every multiple of 22 is not necessarily a refined number. Since we have no other information about the definition of a refined number, we cannot determine how many integers from 1 to 1000 fit that definition.

Statement (2) provides us with a definition of a refined number – a refined number is defined as an even multiple of 11. Since we can find the number of even multiples of 11 in the set, this information is sufficient to answer the question. The answer is B.

# COMPREHENSIVE PROBLEM SET

**25 QUESTIONS, 50 MINUTES**

**Problem 77**                                        **Difficulty level: 2**

Number N is randomly selected from a set of all primes between 10 and 40, inclusive. Number K is selected from a set of all multiples of 5 between 10 and 40, inclusive. What is the probability that N + K is odd?

(A) $\dfrac{1}{2}$

(B) $\dfrac{2}{3}$

(C) $\dfrac{3}{4}$

(D) $\dfrac{4}{7}$

(E) $\dfrac{5}{8}$

## Problem 78

**Difficulty level: 2**

Mark's clothing store uses a bar-code system to identify every item. Each item is marked by a combination of 2 letters followed by 3 digits. Additionally, the three-digit number must be even for male products and odd for female products. If all apparel products start with the letter combination AP, how many male apparel items can be identified with the bar code?

(A)   100

(B)   405

(C)   500

(D)   729

(E)   1000

**Problem 79**                                    **Difficulty level: 2**

Fernando purchased a university meal plan that allows him to have a total of 3 lunches and 3 dinners per week. If the cafeteria is closed on weekends and Fernando always goes home for a dinner on Friday nights, how many options does he have to allocate his meals?

(A)   20

(B)   24

(C)   40

(D)   100

(E)   120

**Problem 80**                                    **Difficulty level: 2**

In a business school case competition, the first three teams receive cash prizes of $5,000, $3,000 and $2,000, respectively, while the remaining teams are not ranked and do not receive any prizes. If there are 6 participating teams, how many outcomes of the competition are possible?

(A)   18

(B)   20

(C)   36

(D)   60

(E)   120

**Problem 81**                                              **Difficulty level: 2**

Jennifer owns 4 shirts of the same design, 2 of which are white, 1 black and 1 red. If all 4 shirts are put on top of each other in a drawer, how many different color arrangements can Jennifer create?

(A)   4

(B)   6

(C)   12

(D)   24

(E)   60

## Problem 82

**Difficulty level: 3**

In how many different ways can a soccer team finish the season with 3 wins, 2 losses and 1 draw?

(A)   6

(B)   20

(C)   60

(D)   120

(E)   240

**Problem 83**                                                                 **Difficulty level: 3**

If a randomly selected non-negative single-digit integer is added to set X {2, 3, 7, 8}, what is the probability that the median of the set will increase while its range will remain the same?

(A)  20%

(B)  30%

(C)  40%

(D)  50%

(E)  60%

## Problem 84

**Difficulty level: 3**

If the President and Vice President must sit next to each other in a row with 4 other members of the Board, how many different seating arrangements are possible?

(A)  120

(B)  240

(C)  300

(D)  360

(E)  720

**Problem 85**                                    **Difficulty level: 3**

What is the number of three-digit multiples of 5 that are not divisible by 10?

(A)  90

(B)  100

(C)  180

(D)  200

(E)  1000

**Problem 86**                                          **Difficulty level: 3**

Jeremy needs to choose 3 flowers for his mother from a group of 10 roses, 6 of which are red and 4 of which are white. What is the ratio of the number of choices Jeremy has to select only red roses to the number of choices he has to select only white roses?

(A)    30:1

(B)    10:1

(C)    5:1

(D)    3:2

(E)    1:5

**Problem 87**                                                    **Difficulty level: 3**

If two elements are dropped from set X {-10, -8, 0, 6, 7}, what will be the percentage change in its mean?

   (1) The median of the set will remain the same
   (2) The range of the set will decrease by 3

(A) Statement (1) alone is sufficient, but statement (2) alone is not sufficient.
(B) Statement (2) alone is sufficient, but statement (1) alone is not sufficient.
(C) BOTH statements TOGETHER are sufficient, but NEITHER statement ALONE is sufficient.
(D) Each statement ALONE is sufficient.
(E) Statements (1) and (2) TOGETHER are NOT sufficient.

**Problem 88**                                      **Difficulty level: 3**

To apply for the position of photographer at a local magazine, Veronica needs to include 3 or 4 of her pictures in an envelope accompanying her application. If she has pre-selected 5 photos representative of her work, how many choices does she have to provide the photos for the magazine?

(A)   5

(B)   10

(C)   12

(D)   15

(E)   50

## Problem 89

**Difficulty level: 3**

Members of a student parliament took a vote on a proposition for a new social event on Fridays. If all the members of the parliament voted either for or against the proposition and if the proposition was accepted in a 5-to-2 vote, in how many ways could the members vote?

(A)   7

(B)   10

(C)   14

(D)   21

(E)   42

## Problem 90          Difficulty level: 3

A retail company needs to set up 5 additional distribution centers that can be located in three cities on the east coast (Boston, New York, and Washington, D.C.), one city in the mid-west (Chicago), and three cities on the west coast (Seattle, San Francisco and Los Angeles). If the company must have 2 distribution centers on each coast and 1 in the mid-west, and only one center can be added in each city, in how many ways can the management allocate the distribution centers?

(A)    3

(B)    9

(C)    18

(D)    20

(E)    36

**Problem 91**

**Difficulty level: 3**

Three couples need to be arranged in a row for a group photo. If the couples cannot be separated, how many different arrangements are possible?

(A)  6

(B)  12

(C)  24

(D)  48

(E)  96

**Problem 92**                                    **Difficulty level: 3**

If 6 fair coins are tossed, how many different coin sequences will have exactly 3 tails, if all tails have to occur in a row?

(A)   4

(B)   8

(C)   16

(D)   20

(E)   24

**Problem 93**                                          **Difficulty level: 3**

If every member of set X {-14, -12, 17, 28, 41, Z} is multiplied by number N, by what percent will the mean M of the set increase?

(1) $Z = 60$

(2) $N = \dfrac{Z}{M}$

(A) Statement (1) alone is sufficient, but statement (2) alone is not sufficient.
(B) Statement (2) alone is sufficient, but statement (1) alone is not sufficient.
(C) BOTH statements TOGETHER are sufficient, but NEITHER statement ALONE is sufficient.
(D) Each statement ALONE is sufficient.
(E) Statements (1) and (2) TOGETHER are NOT sufficient.

**Problem 94**                                    **Difficulty level: 3**

The flag of country Perralia has to contain three stripes of the same width, all of which must be positioned either vertically or horizontally. If the flag of Perralia must consist of the national colors, which include green, red, yellow, black and blue, how many different flags can be created?

(A)   24

(B)   48

(C)   60

(D)   120

(E)   240

**Problem 95**                                                    **Difficulty level: 4**

A telephone company needs to create a set of 3-digit area codes. The company is entitled to use only digits 2, 4 and 5, which can be repeated. If the product of the digits in the area code must be even, how many different codes can be created?

(A)   8

(B)   9

(C)   18

(D)   26

(E)   27

**Problem 96**　　　　　　　　　　　　　　　　　　**Difficulty level: 4**

A group of candidates for 2 analyst positions consists of 6 people. If $\frac{1}{3}$ of the candidates are disqualified and 3 new candidates are recruited to replace them, the number of ways in which the 2 job offers can be allocated will:

(A)　Drop by 40%

(B)　Remain unchanged

(C)　Increase by 20%

(D)　Increase by 40%

(E)　Increase by 60%

**Problem 97**                                          **Difficulty level: 4**

Which of the following could be the range of a set consisting of odd multiples of 7?

(A)   21

(B)   24

(C)   35

(D)   62

(E)   70

## Problem 98                                    **Difficulty level: 4**

Jake, Lena, Fred, John and Inna need to drive home from a corporate reception in an SUV that can seat 7 people. If only Inna or Jake can drive, how many seat allocations are possible?

(A)   30

(B)   42

(C)   120

(D)   360

(E)   720

PROBLEM SET

**Problem 99**                                   **Difficulty level: 4**

What is the probability of selecting a *clean* number from a set of integers containing all multiples of 3 between 1 and 99, inclusive?

(1) A *clean* number is an integer divisible only by 2 factors, one of which is greater than 2
(2) A *clean* number must be odd

(A) Statement (1) alone is sufficient, but statement (2) alone is not sufficient.
(B) Statement (2) alone is sufficient, but statement (1) alone is not sufficient.
(C) BOTH statements TOGETHER are sufficient, but NEITHER statement ALONE is sufficient.
(D) Each statement ALONE is sufficient.
(E) Statements (1) and (2) TOGETHER are NOT sufficient.

**Problem 100**                                    **Difficulty level: 4**

In how many ways can a teacher write an answer key for a mini-quiz that contains 3 true-false questions followed by 2 multiple-choice questions with 4 answer choices each, if the correct answers to all true-false questions cannot be the same?

(A)   62

(B)   64

(C)   96

(D)   126

(E)   128

**Problem 101**                                             **Difficulty level: 4**

A student committee on academic integrity has 90 ways to select a president and vice-president from a group of candidates. The same person cannot be both president and vice-president. How many students are in the group?

(A)   7

(B)   8

(C)   9

(D)   10

(E)   11

## SOLUTIONS

### Answer Key:

| | | | | |
|---|---|---|---|---|
| 77. D | 82. C | 87. B | 92. A | 97. E |
| 78. C | 83. B | 88. D | 93. C | 98. E |
| 79. C | 84. B | 89. D | 94. D | 99. A |
| 80. E | 85. A | 90. B | 95. D | 100. C |
| 81. C | 86. C | 91. D | 96. D | 101. D |

### Solution to Problem 77

Since all primes, except for 2, are odd, number N will always be odd. Then, in order for the sum of N + K to be odd, number K must be even. Thus, our task is to find the number of even multiples of 5 between 10 and 40. Because these numbers have to be divisible by 2 and 5, we will be looking for multiples of 10. The fastest way to compute the probability would probably be to list all multiples of 5 between 10 and 40 and then count the number of multiples of 10:  <u>10</u>, 15, <u>20</u>, 25, <u>30</u>, 35, <u>40</u>. We have 4 multiples of 10 among the 7 possibilities for number K.

Probability (N + K is odd) = Probability (K is even) = $\dfrac{4}{7}$

The answer is D.

### Solution to Problem 78

Since the letter combination is pre-determined, we cannot create any items by changing the letters. Further, since we are dealing with male apparel, the last digit in the 3-digit combination must be even, leaving us with 5 choices for the last digit (0, 2, 4, 6, 8). Because there are no other restrictions, we can use any of the 10 digits (from 0 to 9) in the first and second places of the 3-digit combination.

Thus, we have one combination for the letters, 10 combinations for the first digit, 10 combinations for the second digit, and 5 combinations for the third digit.

Total number of combinations = $1 \cdot 10 \cdot 10 \cdot 5 = 500$

Alternatively, we know there are 1000 combinations of a 3-digit number (think of it as a number lock on a briefcase): 000, 001, 002 ... 997, 998, 999
Since half of these are even, the answer must be 500.

The answer is C.

## Solution to Problem 79

Since the cafeteria is closed on weekends, we have 5 days available for 3 lunches. Since lunch meals are indistinguishable in this scenario, their order does not matter after we the select the lunch days. We can then use the combinations formula to compute the number of choices.

$$\text{Number of choices for lunch} = \frac{5!}{3! \cdot (5-3)!} = 10$$

The similar logic applies to dinners. The only difference is that Fernando prefers to dine at home on Fridays and will therefore have 4 rather than 5 days to use his dinner admission.

$$\text{Number of choices for dinner} = \frac{4!}{3! \cdot (4-3)!} = 4$$

Number of choices for all meals = 4 · 10 = 40

The answer is C.

## Solution to Problem 80

Since only the first 3 teams are ranked in the competition, the number of outcomes will equal the number of ways in which the top 3 teams can be selected from the pool of 6. Since changing the order of the top 3 teams (e.g. switching the first and second places) creates a new outcome, we are dealing with permutations. Thus, we can compute the number of outcomes by using the permutations formula:

$$\text{Number of outcomes} = \frac{6!}{(6-3)!} = 120.$$ The answer is E.

## Solution to Problem 81

To calculate the number of possible color arrangements, we need to find the number of 4-element permutations that can be created from a pool of 4 elements. In other words, we need to find the number of ways in which we can rearrange the 4 colors. Since two of the shirts are of the same color, we have 2 repeating elements and can apply the formula for permutations with repeating elements:

$$\text{Number of color combinations} = \frac{4!}{2!} = 12$$

The answer is C.

## Solution to Problem 82

The easiest approach to this problem is to use what you learned about permutations with repeating elements. In other words, we need to find the number of ways in which we can rearrange the 6 results taking into account that we have 3 repeating elements of wins and 2 repeating elements of losses.

$$\text{Number of ways to end the season} = \frac{6!}{3!\,2!} = 60$$

Alternatively, when we are dealing with multiple selections from a group, it is helpful to consider these selections sequentially, starting with one of the groups and then allocating the remaining elements to other groups. Since switching the games within the "won" group or the "lost" group does not create a new arrangement, we will be using the combinations formula:

$$\text{Number of ways to win 3 games out of 6: } \frac{6!}{3! \cdot (6-3)!} = 20$$

Number of ways to end 1 game in a draw from the remaining 3 = 3

Number of ways to lose 2 games from the remaining 2 = 1
(you can also check this with a general formula: $\frac{2!}{2! \cdot (2-2)!} = \frac{2!}{2! \cdot 0!} = 1$)

Total number of ways to end the season = $20 \cdot 3 \cdot 1 = 60$. The answer is C.


## Solution to Problem 83

Since set X consists of an even number of terms, its median is equal to the average of the two middle numbers, where the elements are arranged in the ascending order:

$$\text{Median of set X} = \frac{3+7}{2} = 5$$

Since after an additional element is added to set X, it will have an odd number of terms, we know that the median of the new set will be the middle element, if the terms of the set are arranged in order. Because we are limited to single-digit integers, the numbers that will increase the median of the set include 6, 7, 8, and 9.

In order for the range of set X to remain the same, the greatest and the smallest values in the set must remain the same. In other words, the new number must be between 2 and 8, inclusive.

By combining both conditions, we know that the numbers that will increase the median without affecting the range are 6, 7, and 8. Since there are 10 non-negative single-digit integers, the probability to select one of these 3 numbers is $\dfrac{3}{10}$ or 30%.

The answer is B.

## Solution to Problem 84

Since the President and Vice President cannot be separated, we can treat them as one unit and treat the other members of the Board as one unit each. Thus, we will have 5 units that will need to be allocated among 5 places. Since the seating patterns can be created only by rearranging the units, the number of such rearrangements is equal to 5! or 120.

Since we can switch the places of the President and Vice President, we will have 2 possibilities for each arrangement of the units.

Total number of arrangements = $2 \cdot 5! = 240$.

The answer is B.

## Solution to Problem 85

Because all multiples of 5 end with a 5 or 0, those of them that end with a 0 will also be divisible by 10. Since we need to exclude the numbers divisible by 10, we need to find the number of 3-digit integers ending with a 5.

Because the first digit in any three-digit number cannot be 0, we have 9 choices for the first digit. The middle digit can be any of the 10 digits, while the last digit has to be 5. Thus, we have 9 choices for the first digit, 10 choices for the second digit and 1 choice for the third digit.

Number of 3-digit numbers fitting the criteria $= 9 \cdot 10 \cdot 1 = 90$

The answer is A.

## Solution to Problem 86

Since the order in which the roses of the same color are selected by Jeremy does not matter, we are dealing with combinations. We can use the combinations formula to compute the number of 3-element selections from a pool of 6 elements (red roses) and 4 elements (white roses).

$$\text{Number of choices for a red-rose bunch} = \frac{6!}{(6-3)! \cdot 3!} = 20$$

Number of choices for a white-rose bunch $= \dfrac{4!}{(4-3)! \cdot 3!} = 4$

Ratio of choices for red roses to white roses $= \dfrac{20}{4} = 5:1$

The answer is C.

## Solution to Problem 87

To find the percentage change in the mean, we need to find the new mean, or more precisely, determine which two elements are dropped. Statement (1) tells us only that the middle number in the set will remain 0. This is insufficient to determine which elements are dropped. For instance, dropping -10 and 7 or -8 and 6, would satisfy the condition in the statement.

From statement (2), we know that the range of the set will change. Thus, the smallest value or the largest value or both of them must be among the elements dropped. Upon closer analysis, you can see that only dropping both the smallest and the largest value will give us the desired decrease in range – from 17 (7 - (-10)) to 14 (6 - (-8)).

Therefore, statement (2) alone is sufficient and the answer is B.

## Solution to Problem 88

Since changing the order in which the photos are put in the envelope does not result in a new group of photos, the order in which the photos are placed does not matter and we are dealing with combinations. Because Veronica can choose to send either 3 or 4 photos, the total number of possibilities can be found by calculating the number of 3-element combinations and the number of 4-element combinations and then adding the two together:

Number of 3-photo selections $= \dfrac{5!}{(5-3)! \cdot 3!} = 10$

Number of 4-photo selections $= \dfrac{5!}{(5-4)! \cdot 4!} = 5$

Total number of possible photo selections $= 10 + 5 = 15$

The answer is D.

## Solution to Problem 89

To answer the question, we need to find the number of ways to divide the 7 members of the parliament into groups of 5 (supporters) and 2 (opponents). Since changing the order of members in each group does not create a new voting pattern (they still remain within supporters or opponents), we can use the combinations formula.

Note that the most efficient way to separate a group into two other unordered groups is to find the number of ways to form the smaller group. After the small group has been selected, all the remaining candidates have to be in the other group.

In other words, to find the total number of voting patterns, we can simply determine the number of ways in which the 2 opponents can be selected from a group of 7. The remaining 5 members will automatically become the supporters of the proposition. Note that this is valid only because the order is not important and hence there is only 1 way to fill the 5 opponent slots with the 5 remaining candidates.

Number of voting patterns = Number of ways to select 2 opponents = $\dfrac{7!}{2! \cdot (7-2)!} = 21$

The answer is D.

## Solution to Problem 90

To simplify the solution of problems involving multiple steps, it is often helpful to consider each of them individually. Let's find the number of choices the company has in each region. Note that the order in which the allocations are made does not matter – i.e. choosing New York and then Boston is the same as choosing Boston and then New York. Thus we will need to use the combinations formula to find the number of possible allocations.

East coast: 2 slots and 3 candidates (Boston, New York and D.C.):

Number of possibilities = $\dfrac{3!}{2! \cdot (3-2)!} = 3$

Midwest: 1 slot and 1 candidate:

Number of possibilities = 1

West coast: 2 slots and 3 candidates (Seattle, San Francisco and L.A.):

Number of possibilities = $\dfrac{3!}{2! \cdot (3-2)!} = 3$

Total number of possible allocations = $3 \cdot 1 \cdot 3 = 9$
The answer is B.

## Solution to Problem 91

Since the couples cannot be separated, we can consider each couple one unit and simply find the number of ways to rearrange the 3 units. The number of rearrangements that can be created with 3 elements is equal 3! or 6.

Note, however, that each couple can also position in 2 different ways: man on the left and woman on the right or vice versa. Since each couple can be rearranged in this way, the number of such rearrangements = $2 \cdot 2 \cdot 2 = 8$.

Total number of rearrangements = $6 \cdot 8 = 48$. The answer is D.

## Solution to Problem 92

Since we need exactly 3 tails, the remaining 3 coins must show heads. You can then view the question as allocating 3 heads and 3 tails to six spots.

Because all the tails have to occur in a row, we can consider them one unit. Since each of the remaining heads can take any spot, we will consider each of the remaining heads a unit. Thus, we have 4 units to allocate among the 4 spots (1 long spot for tails and 3 for heads). Moreover, 3 of the units are exactly the same; heads are indistinguishable from each other. Thus, we need to compute the number of rearrangements of 4 elements, among which 3 are the same:

Number of different sequences = $\dfrac{4!}{3!} = 4$

Note, however, that since this problem is very restrictive and therefore will have few possibilities fitting the description, solving it graphically may be even more time-efficient. While formulas usually have lower chance of leaving out possible outcomes, this particular problem can also be solved quickly by mapping out the sequences fitting the criteria:

(1) TTTHHH     (2) HTTTHH     (3) HHTTTH     (4) HHHTTT

Thus, there are 4 possible sequences and the answer is A.

## Solution to Problem 93

Statement (1) provides us with enough information to compute the original mean M of set X, but provides us with no information about the new mean. Therefore, statement (1) alone is insufficient.

Statement (2) alone is insufficient because we have one equation with two unknowns and have no way of finding either Z or N.

However, plugging the information from statements (1) into statement (2) coupled with the knowledge of the original mean M calculated from statement (1), we can find the value of N and hence the percentage increase of the mean.

The answer is C.

$$\text{Average before} = \frac{-14 - 12 + 17 + 28 + 41 + Z}{6} = \frac{-14 - 12 + 17 + 28 + 41 + 60}{6} = 20$$

$$N = \frac{60}{20} = 3$$

$$\text{Average after} = \frac{3 \cdot (-14) + 3 \cdot (-12) + 3 \cdot 17 + 3 \cdot 28 + 3 \cdot 41 + 3 \cdot 60}{6} = 60$$

Percentage increase = 200%

Note that if we know N, we do not need to know the value of Z to find the new mean:

$$\text{Average before} = \frac{-14 - 12 + 17 + 28 + 41 + Z}{6}$$

$$\text{Average after} = \frac{3 \cdot (-14) + 3 \cdot (-12) + 3 \cdot 17 + 3 \cdot 28 + 3 \cdot 41 + 3 \cdot Z}{6}$$

$$= 3 \cdot \frac{-14 - 12 + 17 + 28 + 41 + Z}{6} = 3 \cdot \text{Average before}$$

If this were not a data sufficiency question, it would be important to note that while the mean after is three times the mean before, the percentage increase is 200%. As an example, think of a company's profit growing by 100% from one year to the next; we say it has doubled – or that it is two times as great.

## Solution to Problem 94

Since each flag has to include 3 of the 5 national colors and the order in which the colors are placed matters, we need to find the number of 3-element ordered arrangements that can be created from a pool of 5 elements. This number will represent the number of possible flags for a determined position of the stripes (either vertical or horizontal). Let's assume that first all the stripes are placed horizontally and find the number of flags with horizontal stripes:

$$\text{Number of flags with horizontal stripes} = \frac{5!}{(5-3)!} = 60$$

Since the stripes can appear either horizontally or vertically, we can create one vertical "twin" for every horizontal color combination simply by changing the horizontal placement into vertical, while keeping the relative position of all colors the same. Since we will be able to create one twin for each horizontal color combination, the opportunity to choose the location of the stripes will double the total number of possible flags:

Number of possible flags with horizontal or vertical stripes = $60 \cdot 2 = 120$

The answer is D.

## Solution to Problem 95

Since the number of elements is small, an efficient way to solve this problem is to find the number of possibilities to fill each of the 3 digit slots and then multiply all the possibilities to get the total number of feasible codes. Furthermore, since there is a constraint on the product of the 3 digits, we will need to reduce the total number of arrangements by the number of possibilities that violate the constraint.

If we did not have the additional constraint on the product of the three digits, we would have 3 possibilities for each of the 3 spots in the area code (each of the 3 digits can be used in any spot because they can be repeated).

Without the constraint, the total number of codes = $3 \cdot 3 \cdot 3 = 27$

Now, we need to subtract the number of possibilities that violate the constraint. Since the product of the 3 numbers will be odd only if all the digits are odd, the only code that violates the constraint is 555. Note that any other code containing at least one 2 or 4 will always yield an even product.

Number of codes yielding an even product = $27 - 1 = 26$

The answer is D.

## Solution to Problem 96

Since the order in which the job offers are extended does not matter, the total number of ways to allocate the offers can be determined using the combinations formula. To find the percentage change in the number of possibilities to allocate job offers, let's compute this number before and after disqualification:

1. Before the disqualification: 6 candidates for 2 slots

$$\text{Number of possibilities} = \frac{6!}{2! \cdot (6-2)!} = 15$$

2. After the disqualification: 7 candidates (i.e. 6-2+3) for 2 slots

$$\text{Number of possibilities} = \frac{7!}{2! \cdot (7-2)!} = 21$$

$$\text{Percentage increase in the number of possibilities} = \frac{21-15}{15} = 40\%$$

The answer is D.

## Solution to Problem 97

Since the difference between two odd numbers must be even, the range of the set must be even, eliminating answers A and C. We also know that since each element in the set is divisible by 7, the difference between any two terms must also be divisible by 7, which eliminates choice B and D.

Thus, we are left with choice E. For example, 70 could be the range of the set if the smallest and largest values were 7 and 77, respectively.

## Solution to Problem 98

Since the driver's seat had better be occupied, we have to allocate the remaining 4 people among the 6 passenger seats. Because the order of the seat assignments is important, we will use the permutations formula:

$$\text{Number of passenger seat arrangements} = \frac{6!}{(4-2)!} = 360$$

This number reflects the possible arrangements for a given driver. Suppose we assigned Jake to drive. Since we can create a twin arrangement for each of the 360 cases by switching the spots of Inna and Jake (asking Inna to drive), the driver choice will double the number of arrangements:

Total seat patterns = 2 · 360 = 720. The answer is E.

## Solution to Problem 99

Statement (1) tells us that a clean number is an odd prime. Since 3 is the only prime number among the multiples of 3 (every other multiple of 3 will be divisible by 1, 3, and itself–more than 2 factors), this statement tells us that there is only 1 clean number among the multiples of 3 less than 100. Since we can find the number of multiples of 3 less than 100, we can find the probability to select number 3 from this set. Thus, statement (1) alone is sufficient.

Statement (2) tells us that a clean number <u>must</u> be odd. Please note that this is just a necessary rather than sufficient condition and it does not mean that any odd number is clean. In other words, if a number is even, it cannot be clean, whereas if the number is odd, it may or may not be clean. Therefore, we do not have any definitive information about the definition of a clean number and cannot find how many clean numbers lie among the multiples of 3 less than 100. Therefore, this statement alone is insufficient. The answer is A.

## Solution to Problem 100

First let's compute the number of possible answer keys without the constraint on the true-false questions. Knowing that there are 2 possibilities for the correct answer on each true/false question and 4 possibilities on each multiple-choice question, we can find the total number of possible answer keys without the constraint:

Number of answer keys without the constraint = $2 \cdot 2 \cdot 2 \cdot 4 \cdot 4 = 128$

Now we need to calculate the number of answer keys that violate the constraint. These answer keys will have all "yes" or all "no" answers to the 3 true/false questions. Note that the answer to the first true/false question can be either "yes" or "no" (2 possibilities) but the remaining answers have to repeat the first, i.e. once the answer to the first question has been determined, we have only one choice for the second and third true/false answers). Since there are no restrictions on the multiple-choice questions, we still have 4 possibilities for each of the 4 multiple-choice questions.

Number of answer keys that violate the constraint = $2 \cdot 1 \cdot 1 \cdot 4 \cdot 4 = 32$

Number of possible answer keys = $128 - 32 = 96$

The answer is C.

### Solution to Problem 101

Let's suppose there are N students in the group of candidates. Since any candidate can be elected president, we have N choices for the first position. Because one of the people has been "used up" after we elect the president, there are N-1 choices to select the vice president. The total number of choices for the two candidates $= N \cdot (N-1)$.

Because we know that there are 90 ways to select the leaders in the committee, we can set up the equation and solve for the number of candidates N in the group:

$$N \cdot (N - 1) = 90$$
$$N^2 - N - 90 = 0$$
$$(N + 9)(N - 10) = 0$$
$$N = 10$$

Therefore, the answer is D.

*Page left intentionally blank.*

PROBLEM SET

# CONCLUDING ADVICE

We hope you have benefited immensely from *Project GMAT*. We believe that it will not only help you maximize your GMAT score, but serve as a useful reference guide at business school too. The concepts we present in this text have a depth that you will better appreciate over multiple readings. You will improve your comprehension of the subjects presented and gain proficiency with the techniques introduced, by reviewing the material again and again.

## FOCUS ON FUNDAMENTALS

To maximize your score on the GMAT, do not rely solely on diagnostic tests and practice problems. Test taking techniques are vitally important, but students who devote the vast majority of their valuable study time to practice tests often repeat the same errors, limiting their potential for improvement. If you encounter a problem that you solved incorrectly, don't be satisfied to simply review the correct answer. Instead, return to the appropriate sections in *Project GMAT* and spend time reviewing the basics. For example, if you are unable to solve a problem about finding the *median* of a set of numbers, it is likely that you will have trouble finding the *range* in another problem.

## BE HUMBLE

Overconfidence serves to undo thousands of test-takers every year. Many students devote too little time to the topics they perceive to be areas of strength. Imagine the math wizard capable of answering any permutation & combination problem thrown at her, but neglecting to review whether 1 is a prime number, or the definition of a "chord" in geometry. Similarly, the former English major might brazenly give short shrift to Sentence Correction, only to stumble over an obscure grammatical rule. The penalty for missing easy problems can even exceed the penalty for missing hard problems as the computer-adaptive nature of the GMAT might place you on a track for a lower score.

## COMPREHENSIVE PREPARATION

Approach the GMAT with a detailed and comprehensive game plan. One-on-one tutoring and preparation courses are ideal for many students. Reputable courses provide structure, proven methods, and proprietary materials. Best of all, you can learn from someone who aced the test and wants to show you how to duplicate his success. The GMAT is more than just a test of one's mental acuity; successful students also possess mental stamina and arrive at the testing center with a meticulous plan for success.

## BEYOND THE GMAT

Regardless of the relative strength of your GMAT score and undergraduate grade point average, admissions committees heavily weigh other factors. To be a competitive applicant at a top school, you will need to submit persuasive and powerful application essays, a time-consuming process. Even subtle differences in essay quality can make the difference between gaining entry into your first choice or settling for a safety school.

If you have any questions about preparing for the GMAT or applying to business school, please contact us. We wish you the best of luck on the entire application process!

*The Veritas Prep Faculty*

CONCLUSION

# VERITAS PREP

### ELITE TEST PREPARATION

In a BusinessWeek survey of more than 6,000 business school graduates from 61 top programs, no area of education received lower marks on average than GMAT prep courses. In response to growing student dissatisfaction with traditional GMAT education, founders Chad Troutwine and Markus Moberg created the Veritas Prep GMAT Course.

From the outset, the founders engineered the Veritas Prep GMAT Course to produce satisfied students and high scores. They opted for the highest possible standards.

### FINEST INSTRUCTORS:

All Veritas Prep instructors undergo thorough training and each has scored in the 99th percentile on an actual GMAT.

### MORE HOURS:

Because the Veritas Prep course is considerably longer than other courses, students have the time to truly master the material.

### SUPERIOR METHODS:

Created by Yale University graduates, the Veritas Prep System is a complete deconstruction of the GMAT combined with proven strategies to attack it with confidence and skill.

### UNIVERSAL SUPPORT:

Seven days a week, Veritas Prep students can call toll-free to seek live help from instructors, or contact them at any time with the 24/7 Online Help system.

### UNRIVALED LEARNING ENVIRONMENT:

Veritas Prep holds all of its classes in exclusive conference facilities and esteemed universities, including Harvard, Yale, Penn, Chicago, and Stanford.

### PERSONALIZED ATTENTION:

Veritas Prep offers more diagnostic feedback than any of its rivals, allowing students an unprecedented opportunity to pinpoint their own weaknesses and address them with Veritas Prep methods.

Competition for entry into graduate business school has grown fierce. Average GMAT scores of admitted students at the top schools continue to rise. Armed with the proven techniques of the Veritas Prep GMAT Course, Veritas Prep students enjoy a competitive advantage over other business school applicants.

To learn more about the services Veritas Prep offers, visit www.veritasprep.com.

CONCLUSION

# VERITAS MBA

### ADMISSIONS CONSULTING

Gaining entry into the world's premier business schools is as competitive as ever. Applicants with impressive work experience and near perfect GMAT scores have no guarantee of admission, while others with comparatively lower stats routinely receive acceptance letters. Admissions committees face the difficult task of selecting from a vast pool of similar candidates. Applicants who know how to effectively market themselves have a tremendous advantage over the competition. Responding to the needs of Veritas GMAT students seeking that advantage, the company founders created the ultimate admissions consulting service. It is now available to everyone.

### COMPREHENSIVE COVERAGE

Our team of experts includes graduates of each of the world's elite business schools, including: Carnegie Mellon, Columbia, Cornell, Chicago, Darden, Duke, Haas, Harvard, INSEAD, Kellogg, LBS, Michigan, MIT, NYU, Stanford, Texas, Tuck, UCLA, Wharton, and Yale.

### INSIDER INFORMATION

All Veritas Admissions Consultants have earned degrees from top tier business schools and formerly, or currently, serve on an MBA admissions committee. Put simply, our team can tell you precisely what each admissions committee is looking for in a successful candidate because they have served on those committees themselves.

### CUSTOMIZED CONSULTING

Rather than offer general, one-size-fits-all advice, Veritas tailors its service to each individual student. Not only do our Admissions Consultants take the time to learn about your unique traits, but we also offer Admissions Specialists in a vast array of categories, including: low GMAT or GPA, older applicant, minority applicant, lower ranked undergraduate, unusual career path, finance /consulting, international, and several other categories.

### FINEST CONSULTANTS

Our experts can demystify the complicated admissions process, offering proven techniques to help you gain admission to your schools of choice. The current team of consultants includes former Assistant Directors of Admissions from Tuck and Columbia, several award-winning writers, and former professional admissions consultants. Collectively our experts have interviewed thousands of MBA candidates, and they have read the applications of tens of thousands more.

To learn more about the services Veritas MBA offers, visit www.veritasmba.com.

CONCLUSION

# NOTES

# NOTES

# NOTES

# NOTES

# NOTES

# NOTES